SpringerBriefs in Electrical and Computer Engineering

More information about this series at http://www.springer.com/series/10059

Ying Wang · Wen'an Zhou
Ping Zhang

QoE Management
in Wireless Networks

 Springer

Ying Wang
Beijing University of Posts
    and Telecommunications
Beijing
China

Ping Zhang
Institute of Network Technology
Beijing University of Posts
    and Telecommunications
Beijing
China

Wen'an Zhou
Beijing University of Posts
    and Telecommunications
Beijing
China

ISSN 2191-8112   ISSN 2191-8120   (electronic)
SpringerBriefs in Electrical and Computer Engineering
ISBN 978-3-319-42452-1   ISBN 978-3-319-42454-5   (eBook)
DOI 10.1007/978-3-319-42454-5

Library of Congress Control Number: 2016946020

Printed on acid-free paper

This Springer imprint is published by Springer Nature
The registered company is Springer International Publishing AG Switzerland

Preface

With the rapid development of technology, there has been a proliferation of new services and applications. The diversity choices of users bring fierce competition as well as huge pressure to service providers. In the meantime, the level of fulfilment of customer demands and user expectations has been the most important indication to distinguish between different providers. Based on this background, the concept of quality of experience (QoE) receives much attention. Moreover, taking advantage of users and services' diversities to smartly design the resource allocation strategy is still one of the most important issues for future wireless networks.

In the past years, we have witnessed rapid progress in the advance of QoE and QoE modeling. However, there still exist some problems. For instance, most literatures on the QoE modelling only focus on the influence of technical parameters and ignore that QoE is multi-dimensional, while the researches emphasizing on various influencing factors do not explain how to model the QoE. Furthermore, how to carry out the QoE management and how to design QoE oriented radio resource management process are still open and challenging tasks. Therefore, this book carries out the study on data-driven QoE management scheme in wireless networks for mobile services.

In Chap. 1, we first give a brief introduction to QoE in wireless communication industry and the necessity to consider user QoE in current mobile service provisioning and transmission. Moreover, personalized QoE management, taking user subjective factors into account, is an emergent topic for refined and better resource utilization. In Chap. 2, QoE definitions are introduced according to different organizations or researchers besides which the state-of-the-art QoE is summarized, including QoE influencing factors, QoE assessment methods, QoE models, QoE management and control applications, and QoE challenges in 5G. To realize personalized QoE management, a data-driven QoE management architecture is proposed in Chap. 3. In Chap. 4, QoE-based resource allocation scheme is studied targeting at QoE maximization. Both conventional non-personalized QoE resource allocation scheme and personalized QoE scheme are presented and a comparison is conducted on simulation results for the two schemes. In Chap. 5, we illustrate how the data-driven QoE assessment is conducted and some experimental details are given. Finally, the concluding remarks are presented in Chap. 6.

The authors would like to thank Peilong Li, Jiajun Liu, Sachula Meng, Qiping Pi, Haiqing Tao, Huan Yu, Yaning Fan, Mengyu Gao, Wenji Zhang, Lijun Song, Yanjun Hou of Beijng University of Posts and Telecommunications, for their contributions in the presented research works.

Beijing, China   Ying Wang
2016   Wen'an Zhou
                                                                                       Ping Zhang

Acknowledgement

This work is supported by National Nature Science Foundation of China (61372113, 61421061).

Contents

1 Introduction 1
   1.1 Mobile Technology Evolution 1
   1.2 Motivation for Personalized QoE Management. 2
   References. 4

2 Background and Literature Survey 7
   2.1 QoE Definition 7
   2.2 Influencing Factors 8
   2.3 Assessment Method 10
       2.3.1 Subjective Assessment 11
       2.3.2 Objective Assessment 11
       2.3.3 Hybrid Assessment 12
   2.4 QoE Models 12
       2.4.1 Mathematic Model 12
       2.4.2 Machine Learning Model 13
   2.5 QoE Management and Control 14
   2.6 Challenges of QoE in 5G. 15
       2.6.1 Challenges from Various Communication Scenarios 16
       2.6.2 Challenges Due to Emerging Applications 16
       2.6.3 Challenges Related to Big Data 17
   2.7 Summary 18
   References. 18

3 Architecture of Data-Driven Personalized QoE Management 21
   3.1 Introduction 21
   3.2 Framework of Data-Driven Personalized QoE Management 22
       3.2.1 Basic Requirements. 22
       3.2.2 Training Module. 22
       3.2.3 Control Module. 25
   3.3 Personalized Character Extraction: User-Service Preference 26
       3.3.1 Bayesian Graphic Model (BGM). 26
       3.3.2 Context Aware Matrix Factorization Model 28

        3.4   Personalized QoE Model and Example User Case   30
        3.5   Summary   32
        References.   32

4   **QoE-Oriented Resource Allocation in Wireless Networks**   33
        4.1   Background   33
                4.1.1   QoS-Based Radio Resource Management Strategies   34
                4.1.2   QoE-Based Radio Resource Management Strategies   34
                4.1.3   Energy Efficiency-Based Radio Resource Management
                            Strategies.   35
        4.2   Traditional QoE-Based Resource Allocation Mechanism   36
                4.2.1   QoE Metric Model   36
                4.2.2   System Model.   38
                4.2.3   Problem Formulation.   39
                4.2.4   Resource Allocation Strategy   39
                4.2.5   Simulation and Analysis   39
        4.3   Personalized QoE-Based Resource Allocation Mechanism   41
        4.4   Summary   43
        References.   43

5   **Implementation and Demonstration of QoE Measurement
        Platform**   45
        5.1   Introduction   45
        5.2   Related Work   45
                5.2.1   Measurement Under Commercial Network
                            Environment   45
                5.2.2   Measurement Under Laboratory Network Environment. . . .   46
                5.2.3   Measurement Under Simulation Network Environment. . . .   47
        5.3   Design of Subjective Measurement.   48
                5.3.1   QoE Related Factors   49
        5.4   Platform Infrastructure on Streaming Media
                Application Scenario.   50
                5.4.1   Supporting System Architecture   50
                5.4.2   Functional Modules.   51
        5.5   Measurement Procedure   52
                5.5.1   Crowdsourcing   52
                5.5.2   Measurement Description   53
                5.5.3   Measurement Result   53
        5.6   Summary   56
        References.   56

6   **Conclusion**   59
        6.1   Conclusion Remarks.   59
        6.2   Future Work   60

Acronyms

ARP	Assignment Reservation Priority
CDMA	Code Division Multiple Access
CLO	Cross Layer Optimization
CMIPC	China Mobile Intellectual Property Center
C-RAN	Cloud-based Radio Access Network
D2D	Device-to-Device
ETSI	European Telecommunications Standards Institute
FEC	Forward Error Correction
GBR	Guaranteed Bit Rate
GSM	Global System for Mobile Communications
HLS	Http Live Streaming
ITU	International Telecommunication Union
LTE	Long Term Evolution
LTE-A	LTE-Advanced
MIMO	Multiple-Input Multiple-Output
MOS	Mean Opinion Score
MPQM	Moving Picture Quality Metric
MR	Measurement Report
MT	Maximum Throughput
OFDMA	Orthogonal Frequency Division Multiple Access
PEP	Packet Error Probability
PF	Proportional Fair
PSNR	Peak Signal to Noise Ratio
PSQA	Pseudo Subjective Quality Assessment
QMF	Quality Management Framework
QoCE	Quality of Customer Experience
QoE	Quality of Experience
QoGE	Quality of Group Experience
QoS	Quality of Service
QoUE	Quality of User Experience

RMSE	Root Mean Squared Error
RNN	Random Neural Network
SMS	Short Messaging Service
SVM	Support Vector Machine
TD-SCDMA	Time Division Synchronous CDMA
TS	Transport Stream
TMF	Tele Management Forum
VNI	Visual Network Index
VQI	Voice Quality Index
VQM	Video Quality Metric
WAP	Wireless Application Protocol
WCDMA	Wideband Code Division Multiple Access
WFL	Weber Fechner Law

Chapter 1
Introduction

Abstract In recent years, with the advancement of wireless communication networks, there is an increasing demand especially on mobile Internet services. Users' Quality of Experience (QoE) becomes one of the main issues for future wireless networks when designing personal and customized services to maintain and attract more users. Furthermore, the research on wireless resource management is moving forward from enhancing objective system performance to improving users' subjective experience. A better QoE-oriented resource allocation policy is preferred and many new challenges are brought out accordingly, including how to quantify and measure QoE, how to design a set of unified wireless resource management strategies and how to make use of a huge amount of available data to derive an optimal QoE model, etc. Therefore, personalized QoE management, efficient estimation, and optimal resource allocation need to be studied and implemented in future wireless networks.

1.1 Mobile Technology Evolution

With the mobile terminal greatly changing our lives over the past decades, the mobile networks and services are playing an increasingly important role and become indispensable for many people. In addition to traditional voice communication service, many new data services are emerging and become popular in the mobile terminal. As the Cisco Visual Network Index (VNI) reported, the number of global mobile users is expected to reach 5.2 billion in 2019 [1].

One reason for this phenomenon is the significant advance in the mobile network technology era. In the last fifty years, the tremendous development of mobile network technologies has been witnessed together with the evolution of mobile services. The Second Generation technologies (2G), e.g., Global System for Mobile communications (GSM) and Code Division Multiple Access (CDMA), have extended the voice-only service to data access service such as Short Messaging Service (SMS) and Wireless Application Protocol (WAP) services. Afterwards, the Third Generation (3G) technologies, e.g., Wideband CDMA (WCDMA) and Time Division Synchronous CDMA (TD-SCDMA), have improved the data access greatly and led

© The Author(s) 2017
Y. Wang et al., *QoE Management in Wireless Networks*, SpringerBriefs
in Electrical and Computer Engineering, DOI 10.1007/978-3-319-42454-5_1

to the variety of mobile multimedia services. The Orthogonal Frequency Division Multiple Access (OFDMA) technology enables the Long Term Evolution (LTE) and LTE Advanced (LTE-A), i.e., the Forth Generation (4G), to provide an even better Quality of Service (QoS) to users with improved data rate. The standardization process for the 4G technologies (LTE, LTE-Advanced) was finished in 2011, and LTE-A networks for business use have been deployed around the world today [2]. In addition, many projects are driven by different countries and organizations around the world for the Fifth Generation (5G) mobile technologies [3, 4]. And in 2015, the relevant testing was launched by both Huawei [5] and Ericsson [6].

1.2   Motivation for Personalized QoE Management

During the development of 5G, it is a consensus that QoE is one of the major issues, considering the user acceptability. In general, QoE is based on the quality of interactions between users and applications, while QoS depends on the quality of interactions between applications and networks. The technologies based on QoE can satisfy the requirements of end users in a better manner than QoS. In the era of 2G, the main service of communication systems is voice service. Thus, the QoS parameters such as Peak Signal to Noise Ratio (PSNR), delay, and coding rate, etc. are well suited to evaluate the quality of communication systems. In 3G and 4G, however, with the popularity of smart phones, there has been a large number of different types of wireless data services, supported in particular by mobile Internet applications. According to the report released by International Telecommunication Union (ITU), by the end of 2014, mobile Internet traffic accounted for 12 % of the total Internet traffic [7]. In China Mobile Intellectual Property Center (CMIPC), top ten applications of mobile Internet are summarized including mobile social, mobile advertise, mobile game, mobile TV, mobile electronic reading, mobile location services, mobile search, mobile content sharing, mobile payment and mobile e-commerce [8]. In addition, a large amount of novel mobile Internet applications are emerging and growing greatly. More complicated parameters are thereafter designed to represent video quality, such as Video Structural Similarity (VSSIM), Video Quality Metric (VQM), and Moving Picture Quality Metric (MPQM).

However, those parameters are still not good enough for the new services, especially when it comes to service context and human subjective factors. In 5G, a greater transmission capacity requirements led to higher carrier frequency, greater bandwidth and larger peak transmission rate. Moreover, the diversities of users within various locations, occupations and economic classes are expected to be concerned in addition to the support of the personalized service. The mobile users tend to pay more attention to their experiences, which leads operators and vendors to provide the products with better user experiences. How to improve the users' QoE of course becomes a key issue when designing customized services. The recent convergence of the mobile Internet has accelerated the demand by changing research direction from original enhancing the system in objective performance to improving user subjective experience. Consequently, in the era of 5G, the more user-oriented parameters are expected to be identified. QoE, as a direct measurement of human perception on the communication service qualities, is promising and beneficial for future communication systems.

The state-of-the-art research work on QoE focuses on how to map between network parameters and allocate resources according to some QoE prediction criteria. Two factors are highly relevant with QoE expectations, including service types and user characteristics. In the meantime, various approaches have been presented and evaluated including fuzzy comprehensive evaluation method, TOPSIS method [9], gray relational analysis [10], neural networks [11], Bayesian networks [12] and contact points [13], etc. In Tele Management Forum (TMF), an end to end Customer Experience (CE) model based on Kilkki model [14] is proposed, and QoE is subdivided into three types including Quality of Customer Experience (QoCE), Quality of User Experience (QoUE) and Quality of Group Experience (QoGE). ITU-T Study Group 12 (2009–2012) proposed Quality Management Framework (QMF) in their Q4/13 [15]. In 3rd Generation Partnership Project (3GPP), the research works on user experience are currently focusing on QoE metric definitions, QoE reporting formats and measurement protocols of QoE negotiation. In addition, since the industrial utility of QoS is pretty mature, it could be a solution to mapping QoE expectation given various service types to QoS parameters such as data rate, delay and packet loss rate, etc. [16] had formulated such mapping as a log linear model. The above research works are also followed by industrial companies and the corresponding infrastructures are established.

In 2010, the Huawei company launched Voice Quality Index (VQI)—National road test program which can monitor the quality of voice, find and locate network quality issues. Given the current network verification, VQI provides abilities to visually identify the network status and the voice quality. Huawei also established a comprehensive network optimization platform named Nastar. Nastar primarily utilizes the measurement report (MR) and call history (CHR) statistics for network analysis and targeting optimization. In the ZTE company, three core user experience management propositions are abstracted including detecting, locating and professional rapid troubleshooting. Although many researches and industrial works are launched on QoE, there are still many issues should to be further explored.

The above works mainly focus on the way to define QoE-oriented parameters. Given those QoE parameters, how to carry out the QoE management is still an open and challenging task. Since the mobile service providers actually do not have unlimited network resources, the high service quality is generally unaffordable by over-provisioning resources. An even allocation strategy is also not suitable due to the diversities of various users and different services, leading to a desperate need of a careful design of the management granularity for each pair of user and service. In order to clarify this issue, we use a soccer video game example illustrated in Fig. 1.1 which has two types of users. One is soccer fan, the other is not. The expectation for service quality of fan is obviously much higher than that of non-soccer fan. Under the limitation of the network resources, a good strategy is to allocate more network resources to fan to obtain a better compromise between user satisfaction and limited resources. Furthermore, an efficient or even real-time solution is essential for estimation and resource allocation. From the above example, both users and services are quite diversified. Therefore, taking advantage of those diversities and smartly

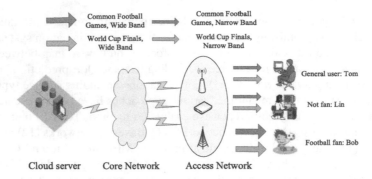

Fig. 1.1 An example of personalized QoE management

designing the resource allocation strategy are beneficial for saving system resource and improving user experience.

In the era of big data, personal user data collection and reservation are feasible under the premise of protecting the human being privacy. In the wireless communication infrastructure, the service data on both mobile terminal and network side can be preserved within a certain period, such as users' cookies stored in the mobile terminals [17] and web log stored in the web server [18]. How to utilize those data becomes an important research issue to optimize QoE. However, the highly diversified data in huge amount makes it impossible to summarize the rules using manual methods. Data-driven approaches originally developed in machine learning area are more promising to trigger an optimal QoE model.

To summarize, currently a huge amount of new services and applications are emerging and growing. How to satisfy the diversified user demanding under the limited wireless resources becomes the critical problem for the service/network providers. Numbers of research issues are prompted on personalized QoE, which is also the key insight within this book. A theoretical background, new architecture for QoE management, QoE-oriented resource allocation and analysis are presented accordingly with practical technical examples.

References

1. T Cisco. Cisco Visual Networking Index: Global Mobile Data Traffic Forecast Update, 2012–2017. *Cisco Public Information*, 2013.
2. 3GPP TS23.402 v10.4.0 architecture enhancements for non-3gpp accesses (release 10). http://www.3gpp.org/DynaReport/23402.htm.
3. A. Osseiran, F. Boccardi, V. Braun, K. Kusume, P. Marsch, M. Maternia, O. Queseth, M. Schellmann, H. Schotten, H. Taoka, et al. "Scenarios for 5G mobile and wireless communications: the vision of the METIS project". *Communications Magazine, IEEE*, 52(5):26–35, 2014.
4. T. Wang, G. Li, J. Ding, Q. Miao, J. Li, and Y. Wang. "5G Spectrum: is china ready?". *Communications Magazine, IEEE*, 53(7):58–65, 2015.

5. Test for 5G in huawei. http://www.huawei.com/ilink/en/abouthuawei/newsroom/pressrelease/HW_329169.
6. Wikipedia. https://en.wikipedia.org/wiki/5G.
7. ITUR Rec. Bt. 500-11, methodology for the subjective assessment of the quality of television pictures. *International Telecommunication Union.*
8. China mobile intellectual property center. http://www.cmipc.org.
9. C. Quadros, E. Cerqueira, A. Santos, and M. Gerla. "A Multi-flow-Driven Mechanism to Support Live Video Streaming on VANETs". In *Computer Networks and Distributed Systems (SBRC), 2014 Brazilian Symposium on*, pages 468–476. IEEE, 2014.
10. O. Markaki, D. Charilas, and D. Nikitopoulos. "Enhancing Quality of Experience in Next Generation Networks Through Network Selection Mechanisms". In *Personal, Indoor and Mobile Radio Communications, 2007. PIMRC 2007. IEEE 18th International Symposium on*, pages 1–5. IEEE, 2007.
11. I. Paudel, J. Pokhrel, B. Wehbi, A. Cavalli, and B. Jouaber. "Estimation of video QoE from MAC parameters in wireless network: A Random Neural Network approach". In *Communications and Information Technologies (ISCIT), 2014 14th International Symposium on*, pages 51–55. IEEE, 2014.
12. K. Mitra, A. Zaslavsky, and C. Ahlund. "Context-aware QoE modelling, measurement, and prediction in mobile computing systems". *Mobile Computing, IEEE Transactions on*, 14(5):920–936, 2015.
13. B. Gardlo, M. Ries, M. Rupp, and R. Jarina. "A QoE evaluation methodology for HD video streaming using social networking". In *Multimedia (ISM), 2011 IEEE International Symposium on*, pages 222–227. IEEE, 2011.
14. TM Forum. Gb962 customer experience management solution suite r15.5.0. *Tele Management Forum.*
15. Itu-t sg 12 - performance, qos and qoe. http://www.itu.int/en/ITU-T/about/groups/Pages/sg12.
16. P. Reichl, S. Egger, R. Schatz, and A. D'Alconzo. "The logarithmic nature of QoE and the role of the Weber-Fechner law in QoE assessment". In *Proceedings of IEEE International Conference on Communications (ICC)*, pages 1–5. IEEE, 2010.
17. L.I. Millett, B. Friedman, and E. Felten. "Cookies and web browser design: toward realizing informed consent online". In *Proceedings of the SIGCHI conference on Human factors in computing systems*, pages 46–52. ACM, 2001.
18. M.C. Burton and J.B. Walther. "The value of web log data in use-based design and testing". *Journal of Computer-Mediated Communication*, 6(3):0–0, 2001.

Chapter 2
Background and Literature Survey

Abstract In this chapter, an overview of QoE is given and the current state-of-the-art background for QoE is described. Specifically, the definitions of QoE from different aspects and QoE influencing factors are first presented. Then, QoE assessment methods and QoE models are introduced in the following section. In addition, QoE management and control issues are also investigated. Last but not least the challenges of QoE in 5G wireless networks are discussed.

2.1 QoE Definition

As far as we know, there is no common sense about how to define QoE. Given the description of Wikipedia, QoE is considered as a subjective measure of customer's experiences with a service [1]. That means different types of QoE are involved in various services. For example, when a user shops online, QoE is associated with the process of purchasing and the service. When a user has a meal or haircut, what involves QoE is the process of user being in service.

In this book, we discuss QoE in the scenes of wireless communications. For example, users call others using a mobile phone or browse the web using a tablet computer or watch online video via a notebook. In the process of being in service, users get certain service experiences. In such a scenario, a number of international organizations proposed the following definitions:

According to ITU, QoE is described as *the overall acceptability of an application or service, as perceived subjectively by the end user* [2], while the European Telecommunications Standards Institute (ETSI) defines QoE as a measure of user performance based on both objective and subjective psychological measures of using an ICT service or product [3]. Recently, based on the results of the research project COST Action IC1003, EU Qualinet Community proposes the definitions of QoE as *the degree of delight or annoyance of the user of an application or service.* In the context of communication services, QoE depends on the service, content, device, application, and context of use [4].

In addition, different researchers also give different definitions of QoE according to their own understanding. In [5], Kilkki broadens the concept of QoE to the basic

© The Author(s) 2017

Y. Wang et al., *QoE Management in Wireless Networks*, SpringerBriefs
in Electrical and Computer Engineering, DOI 10.1007/978-3-319-42454-5_2

character or nature of direct personal participation or observation. Khalil believes ITU definition of QoE is hard to involve objective human factors. Therefore, he puts forward his own definition: *a blueprint of all human subjective and objective quality needs and experiences arising from the interaction of a person with technology and business entities in a particular context* [6]. Pyykko presents the definition of QoE in the manner of a mobile video scenario: *the binary measure to locate the threshold of minimum acceptable quality that fulfills user quality expectations and needs for a certain application or system* [7].

Although there is no universal answer for what QoE is, we can draw the conclusion that: QoE is a kind of user satisfaction description during the process of interactions between users and services. That is, users get some subjective feelings which is assessed by QoE.

2.2 Influencing Factors

QoE is a multidisciplinary research topic based on social psychology, cognitive science, economics, engineering science, etc. Researchers always limited QoE only in some situations which they were concerned about, making it difficult to understand QoE factors from a global perspective. It is indicated that the video codec parameters along with user expectations should be considered when evaluating the video services and that the context factors such as actual environmental context and social cultural context also play a critical role [8, 9]. T. Hoßfeld et al. of [10] discover that current page loading time as well as history page loading time has impact on user QoE for web browsing services, which can be summarized as memory effect that current experience is highly related with history experience. A conclusion is drawn in [11] that user's mental mood, context, characteristics of mobile devices, network access capacity will all influence user QoE to different extents. For a better and more comprehensive understanding of QoE influencing factors, K. Kilkki and K.U.R. Laghari consider QoE from the perspective of communication ecosystem [6]. Communication ecosystem is defined as the systematic interaction of living (human) and non-living (technology, and business) in a particular context [5], which is first put forward by K. Kilkki and then refined by K.U.R. Laghari. In this ecosystem, QoE influencing factors can be divided into four domains: human domain, technology domain, business domain and context domain. A summary of QoE factors studied in existed literatures is given in Table 2.1.

However, it is still a challenge to manage these QoE influencing factors. First, it is difficult to quantize some of these factors which mainly include subjective factors such as interests, expectations, etc. It is also hard to analyze and quantize the relationship of QoE with respect to these factors. Second, some of these factors such as some contextual factors (quiet or noisy) are intricate to be obtained in the real application scenarios. Thirdly, it is difficult to accurately model the relationship between these factors and QoE, and we should research more on how those factors influence QoE synthetically. Hence, further study is needed to analyze QoE influencing factors.

Table 2.1 Summary of QoE factors

	Application scenario	Factors considered	Details
[12]	Not given	Technical performance	Application, service, network, device performances
		Usability	Application, network, device usability
		Subjective assessment	Application, network, device assessment
		User expectation	Decided by application types, user personality and history experiences
		Context	Physical, social, technical, cultural context
[13]	Video, VoIP, multi-player game	Application dimension	Content type, video codec resolution, frame rate
		Resource dimension	Network resource parameters (delay, jitter, packet loss rate), hardware resource parameters (memory, screen resolution)
		Context dimension	Light, position, time, social environments
		User dimension	Individual property, preference, expectation, requirement and mental mood
[4]	Multi-media services	Human dimension	Human property, social and economical background, physical and psychological status
		System dimension	Content parameters (content type), media parameters (encoding, sampling rate, frame rate), network parameters (bandwidth, delay, jitter), device parameters (memory, screen resolution)
		Context dimension	Physical context (position, time), economical context (cost, brand image), task context (single or multi task)

(continued)

Table 2.1 (continued)

	Application scenario	Factors considered	Details
[6]	Communication systems	Technical domain	Service factors (service content), network factors (delay, jitter, throughput), device factors (device feature, function)
		Business domain	Consumer business model factors (advertisement, brand effect), inter-company business model (commercial strategy), intra-company business model factors (service level agreement)
		Human domain	Population property (age, gender), role factor (consumer, user)
		Context domain	Real context (time, position, climate), virtual context, social context (social relationships)
[14]	Mobile video	Human domain	Population attribute, requirements, mood, expectations, motivations
		System domain	Device factor (memory, battery life, screen size), network factor (network bandwidth, channel condition), video provider factor (encoding algorithm, business model)
		Context domain	Physical context (position, time), social context (alone or in groups), task context (single or multi task)

2.3 Assessment Method

In general, QoE evaluation methods can be divided into three categories, that is, subjective assessment, objective assessment and hybrid assessment. The detailed descriptions are as follows.

2.3.1 Subjective Assessment

Subjective evaluation method is conducted based on psychological/visual experiments, which is the most reliable but also the most complicated and expensive method of evaluating user' QoE. It has been researched for many years, giving researchers deeper understandings of QoE subjective dimension. Most of the output of the subjective evaluation experiment is the opinion score when the user is being served or has been served, and these scores are ultimately averaged into Mean Opinion Score (MOS) [15]. Meanwhile, ITU also sets some corresponding standards of conducting the experiment. The QoE evaluation for video services is the most complicated and some standards are set to conduct the experiment of evaluating subjective video quality. For example [16], the different experiment settings include Single Stimulus (SS), Double Stimulus Impairment Scale (DSIS), Double Stimulus Continuous Quality Scale (DSCQS), Single Stimulus Continuous Quality Evaluation (SSCQE), Simultaneous Double Stimulus for Continuous Evaluation (SDSCE), and Stimulus Comparison Adjectival Categorical Judgment (SCACJ). These settings are similar and the changes mainly reflect in metrics, video reference, video length, number of users, number of observers, etc.

The results of subjective evaluation method are the most accurate, owing to a direct gain of data from the users. Nonetheless, the human cost of this method is too high and cannot be used to automation and real-time situation.

2.3.2 Objective Assessment

Objective evaluation method is defined as using separately the measurement of objective quality to evaluate the subjective quality [17]. In other words, objective evaluation method provides a mapping model from the objective quality to the subjective quality. A variety of objective quality evaluation and prediction models have been studied. Each model has its applicable scenarios and corresponding constraints.

There are many objective evaluation methods to assess QoE which can be generally classified into three kinds: full reference, no reference and partial reference [18]. We take the video business for example to illustrate the difference of these three categories. The full reference method is to compare reference video and test video frame by frame while no reference method is to analyze test video only without reference video. Partial reference method is the compromise of the first two which extracts some characteristics from the reference video and then analyzes the test video according to these characteristics.

The advantage of an objective evaluation method lies in its convenience and tractability. Researchers only need concern about QoS parameters which can be measured and related mathematical problems. It also has its disadvantage of inaccuracy, i.e., the QoE obtained is only an approximation rather than a precise value for each user.

2.3.3 Hybrid Assessment

Except the above two grading methods, there is another method of mixing both and evaluating users' QoE more accurately which is Pseudo Subjective Evaluation Method (PSQA) [19]. This method is a kind of method which is based on statistical learning and uses the Random Neural Network (RNN). In its evaluation, it usually needs to go through four stages:

- Generating influencing factors. At this stage, a lot of service samples after parameter optimization will be generated in the database.
- Measuring the subjective quality. At this stage, testers of the experiment will grade service samples of step 1.
- Training the neural network model. This is the core of this approach. Data collected by the former two steps will constitute the training data set and is used to train the random neural network. The random neural network can be substituted by other machine learning tools, such as the Bayesian network, the Support Vector Machine (SVM).
- Using the neural network model for evaluation. After the third step, a trained RNN will be obtained. So in the evaluation, it is only needed to put the target service data into RNN. Then it can calculate the corresponding QoE of the user.

PSQA is a kind of widely used QoE evaluation methods, applied in [20–26].

2.4 QoE Models

2.4.1 Mathematic Model

Mathematical models are one of the traditional ways which map the influencing factors to the users' QoE. The data of related factors and users' QoE are obtained in a laboratory environment in general. Researchers conduct statistical analysis to formulate the specific relationship between QoE and the parameters. Some models which belong to this type are depicted below.

The E-model recommended by ITU-T SG12 is a classic linear model which is used to predict the overall quality in a voice conversation at the network planning stage. Although E-model is a mature one, it is restricted in voice service over telecommunication networks.

Peter Reichl argues that QoE has a logarithmic nature based on the Weber-Fechner Law (WFL) [27]. The basic idea of WFL is that the human sensory system can be traced back to the percipience of so-called "just noticeable differences" and the differential perception is directly proportional to the relative change of physical stimulus. By taking network QoS as stimulus and QoE as the perception, we will obtain a logarithmic relationship mapping from QoS to QoE. Although the logarithmic model is very convincing which lies in that it is based on the psychological theory, there are limitations that the input QoS should be viewed as a physical stimulus. However, the fact is that many factors could not be regarded as physical stimulus. Furthermore, this model only processes limited numbers of factors, which indicates that it cannot be extended.

Another mathematical QoE model is based on the "IQX hypothesis" in [28], which argues that a change of QoE depends on both the identical QoS changes and the actual level of the QoE. After the integration of the formulation of QoE and QoS, a negative exponential function of QoE w.r.t QoS impairment factors such as packet loss can be acquired. It is found that QoE relates to the QoS impairment factors such as packet loss or network delay in the model of IQX hypothesis, while QoE is related to the perceivable QoS resource like bandwidth or bit rate in the logarithmic model.

2.4.2   Machine Learning Model

In recent years, researchers find out that it is hard to formulate the relationship between influencing factors and users' QoE explicitly by a mathematic model in most cases, not to mention the unobtainable system parameters. Machine learning methods are widely applied to solve the problem of the connotative relationship between QoE model and the influencing factors.

The Recurrent Neural Network (RNN) model is a classical one of machine learning models which is applied in the Pseudo Subjective Quality Assessment (PSQA) assessment. RNN is made up of a group of neurons which can communicate with each other by signals. In the RNN network, the state of every neuron is a nonnegative integer named potential and it can be changed by the signals coming from other neurons. At the end of the training, every neuron shall have a computed potential. Accordingly, the QoE assessment will be obtained by synthesizing the potentials of the RNN.

In [29], T. Hoßfeld et al. uses Support Vector Machine (SVM) method to assess QoE. A hyperplane can be calculated according to training set data and validated with test data set, then the trained SVM model will be applied to evaluate user current QoE with input factors including current page loading time and user history ratings.

Besides, Decision Tree (DT), another machine learning model, can also be applied to build QoE model. DT is a widely used classification model and the relationship between QoE and researching influencing factors can be learned by decision tree building and pruning as in [30]. In [30], a training data set with input (continuous parameters including time, spatial, bit rate, frame rate information) and output (binary values indicating whether current quality is acceptable or unacceptable for users) is utilized to train the tree model after some pre-processing measures. QoE model established in this way can then be used to predict user's attitude towards the service and [30] illustrates that QoE prediction by DT is more precise than that by SVM.

To sum up various QoE models, a comparison for QoE modeling approach is given in Table 2.2.

Table 2.2 Comparison for QoE modeling approach

Method	Type of service	Factors	Metric	Precision	Complexity
Logarithmic function [31]	File downloading	Bandwidth, file size	MOS	RMSE = 0.063	Low
Exponential function [32]	VoIP	Packet Loss	MOS	R = 0.998	Low
		Rearrangement	MOS	R = 0.993	
	Web browing	Waiting time	MOS	R = 0.966	
		Bandwidth	MOS	R = 0.951	
SVM (Support Vector Machine) [30]	Video (Cellphone, Tablet PC, PC)	Time, space, bit rate, frame rate	Acceptability (Binary)	Precision 88.592.85 % 89.382.77 % 91.452.66 %	Medium
Decision tree [30]	Video (Cellphone, tablet PC, PC)	Time, space, bit rate, frame rate	Acceptability (Binary)	Precision 93.551.76 % 90.292.61 % 95.462.09 %	Medium
Random neutral network [24]	Video	Packet loss, packet loss time	MOS	Unaccounted	High

2.5   QoE Management and Control

Based on the QoE models and real-time QoE monitoring, intelligent QoE management can be further conducted by network operators based on actual network conditions. Appropriate measures are taken according to the network problem and optimization methods. On one aspect, quality improvement of a current flow, or maximization of system average QoE can be achieved by reasonable network controls, e.g. admission control, priority decision, congestion control and packet scheduling. A common used way to optimize QoE is to perform cross layer optimization (CLO). For example, in [28], a cross-layer model is established for multimedia traffic in mobile communication systems. QoE is obtained by mapping from the network layer parameter symbol rate. The maximized QoE can be achieved by adjusting the symbol rate for each user with a greedy algorithm. A wise resource allocation strategy with QoE awareness can be realized to efficiently save network resources without user experience degradation.

Similar optimization scheme for wireless video transmission is provided in [21] for stream transmission. First a video stream is divided into two sub-streams using multiple description coding scheme and then two paths with the highest available bandwidth are selected by the server to send these two streams to the client. User QoE is assessed by PSQA based on packet loss rate when two sub-streams arrive at the client. The server will make a decision to redirect the two sub-streams over new paths with higher available bandwidth if the assessed QoE cannot meet the users' requirement.

Selecting a proper access point is another way to improve QoE. In [33], Kandaraj presents a user-based and network assisted scheme for network selection in wireless IEEE 802.11 technology. In the scheme, a PSQA tool is implemented to evaluate users QoE for each access points (AP). When a new user comes, he or she will get a score based on the evaluated QoE and status from each AP. So a better quality of connection can be chosen, resulting in a higher QoE for the user.

In some literatures, researchers consider QoE optimization from a system architecture level. Latre et al. in [34] proposed a QoE-based network management architecture in multimedia services with a three-plane approach (a monitor plane to monitor network parameters, a knowledge plane to determine the optimal QoE decisions and an action plane to conduct decided actions). Optimization actions include applying Forward Error Correction (FEC) to reduce the packet loss caused by errors on a link and switching to different video bit rate to avoid congestion or to obtain a better video quality.

Additionally, QoE-driven network access selection control, handover decision control and rate adaptation control are also studied in many literatures as [33, 35, 36].

2.6 Challenges of QoE in 5G

In the era of 5G, a fiber-like access data rate and a "zero" latency user experience is expected to be provided. Delivering a consistent experience is required across a variety of scenarios including the cases of ultra-high traffic volume density, ultra-high connection density, and ultra-high mobility. 5G will also be able to provide ability on intelligent optimization based on services and users awareness, and improve energy and cost efficiency by over a hundred of times. These expectations above all inspire us to achieve the vision of 5G—"Information a finger away, everything in touch."

To realize these expectations, several new technologies, e.g., heterogeneous networks, millimeter-wave (mm-wave), device-to-device (D2D)/machine-to-machine (M2M), cloud-based radio access network (C-RAN), and smart devices are addressed in the 5G network to cater to the notable trends of 5G, such as explosive growth of data traffic, massive increase in the number of interconnected devices, continuous emergence of new services and application scenarios, etc., which will inevitably impact user QoE, positively or negatively. New challenges are raised for QoE management in 5G and better QoE performance is capable to be achieved if these issues are well handled.

2.6.1   Challenges from Various Communication Scenarios

In 5G, the scenarios of communications are greatly broadened, typically, e.g., when people are working or at entertainment, stationary or moving, at home or in stadiums, subways, highways. For the various scenarios above all, a consistent QoE is desired to achieve for end users. However, QoE management is significant but challenging in the high complicated scenario including ultra-high traffic volume density, ultra-high connection density, and/or ultra-high mobility, etc. For example, it is important to maintain a satisfactory level of service continuity in high speed scenario. Similarly, improving QoE is also important in the ultra-dense small cells scenarios. Those requirements are demanding more appropriate mobility management strategies. Therefore, how to control QoE to satisfy the various requirements in the various scenarios needs to be comprehensively investigated accordingly. A QoE-based dynamic resource allocation for mobile cloud computing environment to satisfy the requirements of end users is proposed in [37]. In [38], a QoE-based resource allocation problem is pointed out in a D2D scenario, where user QoE is evaluated from the both perspectives of stall events and video qualities. While key QoE influencing factors and user expectations may vary for distinct scenarios, QoE becomes a comprehensive metric and an important topic about how to provide personalized user experience through a more refined resource management.

2.6.2   Challenges Due to Emerging Applications

The cellular technology has changed the way we communicate and our social life [39]. Over the past era of 2G, phone call and SMS are the main functions of the mobile terminals. Today 3G/4G become mainstream and the smart terminals are so popular that makes the applications more diverse, such as mobile game, mobile music, mobile video, etc. In the future of 5G, more stunning new technologies and applications will be integrated and enter into human life such as virtual reality, 3D videos etc., which accordingly leads to two challenges for QoE research.

One challenge is the demand to develop appropriate and reasonable QoE models for the emerging applications. Due to the rapid increase of intelligent terminal on capacity and performance, there are a large number of new applications emerging [40]. For example, unlike traditional terminals, the new smart wearable terminals have various forms, such as smart watches, smart glasses, sports wristbands, smart jewelry, etc. Meanwhile, mobile services, which are based on the intelligent terminals, are no longer limited to telephone, text messaging and video services, but involved in medical monitoring, interactive games, information exchange and other emerging businesses. However, the state-of-the-art research works on QoE focus on VoIP, video, and HTTP services. Further study is needed for the new developed applications in order to build up a proper QoE model.

Imagine a medical monitoring service running on a wearable equipment, which monitors blood pressure, displays status on pulse and even identifies whether the user has an irregular heartbeat. In this case, the QoE model should utilize the individual user's information to make accurate decisions to achieve a high level of QoE. For example, for a user with hypertension according to the historical records, when symptoms appear, the system is required to ensure realtime transmission and a certain level of reliability to activate the emergency alarm and other corresponding medical-related services. It is essential to combine the various factors to guarantee QoE.

The smart mobile terminals for human life provide a great convenience, so that people are increasingly dependent on their smartphones. Consequently, the people are increasingly sensitive to the life of the battery. If a service consumes too much power, the users will have very bad feelings, which leads to a lower QoE degree. As a result, the energy efficiency is also an important QoE factor in 5G.

2.6.3   Challenges Related to Big Data

In the 5G era, due to the network data traffic explosion and applications diversification, a huge amount of data is expected to be transfered during the mobile communications [41], which leads to some new challenges in the QoE management field.

One challenge concerning big data is the subjective QoE influencing factors. The main research issue on this topic includes two aspects. The first is the subjective QoE factors including a user specific preference, users' mood, attention, expectation etc. The factors above all are qualitative and highly related with social psychology and cognitive science fields. In order to use these factors in practical applications, some problems should be concerned including: (1) how to quantize these factors? (2) how to normalize these factors after quantization? and (3) how to capture and monitor these factors in a practical real-time application scenario. In addition, it is difficult for a certain application/service to decide which factors are essential and which are not. How to smartly model QoE given a service/application is quite important issue for further investigations.

Another challenge is about QoE evaluation from the perspective of big data. Most existing research on subjective QoE evaluation is conducted based on users' direct feedback of the service. It is suitable for a relative small scale of evaluation. When more data is needed which makes the results more reliable, it is too expensive and even unaffordable by using the above evaluation ways. With the development of big data technologies, many user specific information is preserved and can be used to infer the user's subjective feelings, which is the basis of QoE. As an example of online video services, the viewing time, the total number of clicks on a certain video, the number of daily accesses etc. could be measured and analyzed for user's QoE. For Web services, the online reviews is the knowledge resources in which the relevant features could be extracted reflecting user experience feedback on the web services. Many similar research works had been carried out in the manner of the data-driven way in other areas such as natural language processing (NLP) and unstructured data processing, Thus, it is meaningful to use data-driven approaches in the topic of mobile QoE in the future.

It is also challenging to handle the security and privacy related to intelligent terminals, which comes along with the trend that intelligent terminals play a key role in 5G, especially in the big data era. For those who use intelligent terminals, intentionally or unintentionally, their personal information such as contacts, download histories, application usage records, the system logs, etc., are saved in either client end or cloud end. It is of course beneficial to make use of those data to facilitate QoE. But those data are possibly used for the wrong purpose, even illegal use. How to balance the protection requirement of user's privacy and the QoE management usage on the personalized data is one of the major challenges.

2.7 Summary

In this chapter, an overview for the state-of-the-art QoE is given, including QoE influencing factors, QoE assessment methods, QoE models, QoE management and control applications, and QoE challenges on 5G. QoE definitions are summarized at first. The QoE influencing factors are then discussed in a systematic way, followed by the sequential part which QoE assessment methods are introduced and QoE models are summarized. QoE management and control issues are also discussed in this chapter. Furthermore, the challenges of QoE in 5G wireless networks are discussed finally.

References

1. Quality of experience. http://www.wikepedia.org.
2. ITU Telecommunication Standardization Sector and OF ITU. "Definition of quality of experience (QoE), liaison statement". *ITU-T Recommendation P.10/G.100, Amd 1*, 2007.
3. European Telecommunication Standards Institute TC HF (Human Factors). "Quality of Experience (QoE) requirements for real-time communication services". Technical report, European Telecommunication Standards Institute, 2010.
4. P. Le Callet, S. Moller, and A. Perkis. "Qualinet white paper on definitions of quality of experience (QoE)", 2013.
5. K. Kilkki. "Quality of Experience in Communications Ecosystem". *J. UCS*, 14(5):615–624, 2008.
6. K. U. R. Laghari and K. Connelly. "Toward total quality of experience: A QoE model in a communication ecosystem". *Communications Magazine, IEEE*, 50(4):58–65, 2012.
7. S. Jumisko-Pyykkö, V.K. Malamal Vadakital, and M.M. Hannuksela. "Acceptance threshold: A bidimensional research method for user-oriented quality evaluation studies". *International Journal of Digital Multimedia Broadcasting*, 2008.
8. H. Rifai, S. Mohammed, and A. Mellouk. "A brief synthesis of QoS-QoE methodologies". In *Programming and Systems (ISPS), 2011 10th International Symposium on*, pages 32–38. IEEE, 2011.

9. S. Möller, A. Raake, M. Wältermann, and N. Côte. "Towards a universal scale for perceptual value". In *Quality of Multimedia Experience (QoMEX), 2010 Second International Workshop on*, pages 142–146. IEEE, 2010.

10. T. Hoßfeld, S. Biedermann, R. Schatz, A. Platzer, S. Egger, and M. Fiedler. "The memory effect and its implications on Web QoE modeling". In *Proceedings of the 23rd International Teletraffic Congress*, pages 103–110. International Teletraffic Congress, 2011.

11. S. Ickin, K. Wac, M. Fiedler, L. Janowski, et al. "Factors influencing quality of experience of commonly used mobile applications". *Communications Magazine, IEEE*, 50(4):48–56, 2012.

12. S. Baraković, J. Baraković, and H. Bajrić. "Qoe dimensions and QoE measurement of NGN services". In *in Proc. Telecommunications Forum*, 2010.

13. L. Skorin-Kapov and M. Varela. "A multi-dimensional view of QoE: the ARCU model". In *in Proc. International Convention MIPRO*, pages 662–666. IEEE, 2012.

14. W. Song, D. Tjondronegoro, and M. Docherty. *"Understanding user experience of mobile video: framework, measurement, and optimization"*. INTECH Open Access Publisher, 2012.

15. ITUT Rec. "P. 800.1, mean opinion score (mos) terminology". *International Telecommunication Union, Geneva*, 2006.

16. I. Rec. "Bt. 500-11, methodology for the subjective assessment of the quality of television pictures". *International Telecommunication Union*, 2002.

17. W. Cherif, A. Ksentini, D. Négru, and M. Sidibé. "A_PSQA: Efficient real-time video streaming QoE tool in a future media internet context". In *Proceedings of 2011 IEEE International Conference on Multimedia and Expo (ICME)*, pages 1–6. IEEE, 2011.

18. Fernando Kuipers, Robert Kooij, Danny De Vleeschauwer, and Kjell Brunnström. *"A Techniques for measuring quality of experience"*. Springer, 2010.

19. G. Rubino. "The PSQA project". *INRIA Rennes*, http://www.irisa.fr/armor/lesmembres/Rubino/myPages/psqa.html, 2010.

20. M. Ghareeb and C. Viho. "A multiple description coding approach for overlay multipath video streaming based on QoE evaluations". In *Proceedings of International Conference on Multimedia Information Networking and Security (MINES)*, pages 39–43. IEEE, 2010.

21. M. Ghareeb and C. Viho. "Hybrid qoe assessment is well-suited for multiple description coding video streaming in overlay networks". In *Proceedings of 2010 Eighth AnnualCommunicati on Networks and Services Research Conference (CNSR)*, pages 327–333. IEEE, 2010.

22. K. Piamrat, A. Ksentini, J. Bonnin, and C. Viho. "Q-DRAM: QoE-based dynamic rate adaptation mechanism for multicast in wireless networks". In *Proceedings of IEEE Global Telecommunications Conference*, pages 1–6. IEEE, 2009.

23. X. Sun, K. Piamrat, and C. Viho. "QoE-based dynamic resource allocation for multimedia traffic in IEEE 802.11 wireless networks". In *Proceedings of 2011 IEEE International Conference on Multimedia and Expo (ICME)*, pages 1–6. IEEE, 2011.

24. K. Piamrat, C. Viho, J. Bonnin, and A. Ksentini. "Quality of experience measurements for video streaming over wireless networks". In *Proceedings of 2009 Sixth International Conference on Information Technology: New Generations*, pages 1184–1189. IEEE, 2009.

25. K. Piamrat, K. D. Singh, A. Ksentini, C. Viho, et al. "QoE-aware scheduling for video-streaming in high speed downlink packet access". In *Proceedings of 2010 IEEE Wireless Communications and Networking Conference (WCNC)*, pages 1–6. IEEE, 2010.

26. P. Reichl, B. Tuffin, and R. Schatz. "Economics of logarithmic quality-of-experience in communication networks". In *Proceedings of 2010 9th Conference on Telecommunications Internet and Media Techno Economics (CTTE)*, pages 1–8. IEEE, 2010.

27. P. Reichl, S. Egger, R. Schatz, and A. D'Alconzo. "The logarithmic nature of QoE and the role of the Weber–Fechner law in QoE assessment". In *Proceedings of IEEE International Conference on Communications (ICC)*, pages 1–5. IEEE, 2010.

28. S. Khan, S. Duhovnikov, E. Steinbach, and W. Kellerer. "MOS-based multiuser multiapplication cross-layer optimization for mobile multimedia communication". *Advances in Multimedia*, 2007(1):1–11, 2007.

29. T. Hosfeld, S. Biedermann, R. Schatz, A. Platzer, S. Egger, and M. Fiedler. "The memory effect and its implications on Web QoE modeling". In *Proceedings of 2011 23rd International Teletraffic Congress (ITC)*, pages 103–110, 2011.

30. V. Menkovski, A. Oredope, A. Liotta, and A. C. Sánchez. "Predicting quality of experience in multimedia streaming". In *in Proc. the 7th International Conference on Advances in Mobile Computing and Multimedia*, pages 52–59. ACM, 2009.
31. P. Reichl, S. Egger, R. Schatz, and A. D'Alconzo. "The logarithmic nature of QoE and the role of the Weber–Fechner Law in QoE assessment". In *Proceedings of 2010 IEEE International Conference on Communications (ICC)*, pages 1–5. IEEE, 2010.
32. M. Fiedler, T. Hossfeld, and P. Tran-Gia. "A generic quantitative relationship between quality of experience and quality of service". *IEEE Network*, 24(2):36–41, 2010.
33. Kandaraj Piamrat, Adlen Ksentini, César Viho, and Jean-Marie Bonnin. "AqoE-aware admission control for multimedia applications in ieee 802.11 wireless networks". In *Vehicular Technology Conference, 2008. VTC 2008-Fall. IEEE 68th*, pages 1–5. IEEE, 2008.
34. S. Latré, P. Simoens, B. De Vleeschauwer, et al. "An autonomic architecture for optimizing QoE in multimedia access networks". *Computer Networks*, 53(10):1587–1602, 2009.
35. N. Amram, B. Fu, G. Kunzmann, T. Melia, D. Munaretto, S. Randriamasy, B. Sayadi, J. Widmer, and M. Zorzi. "QoE-based transport optimization for video delivery over next generation cellular networks". In *in Proc. IEEE Symposium on Computers and Communications (ISCC)*, pages 19–24. IEEE, 2011.
36. S. Thakolsri, W. Kellerer, and E. Steinbach. "QoE-based cross-layer optimization of wireless video with unperceivable temporal video quality fluctuation". In *in Proc. IEEE International Conference on Communications (ICC)*, pages 1–6. IEEE, 2011.
37. W. Ying, M. Sachula, C. Yongce, S. Ruijin, Xinshui W., and Kai S. "Multi-leader multi-follower stackelberg game based dynamic resource allocation for mobile cloud computing environment". *Wireless Personal Communication, to appear*.
38. Hao Zhu, Yang Cao, Wei Wang, Boxi Liu, and Tao Jiang. "Qoe-aware resource allocation for adaptive device-to-device video streaming". *Network, IEEE*, 29(6):6–12, 2015.
39. G. Fettweis and S. Alamouti. "5G: Personal mobile internet beyond what cellular did to telephony". *IEEE Communications Magazine*, 52(2):140–145, 2014.
40. B. Bangerter, S. Talwar, R. Arefi, and K. Stewart. "Networks and devices for the 5G era". *IEEE Communications Magazine*, 52(2):90–96, 2014.
41. W. Hummer, S. Schulte, P. Hoenisch, and S. Dustdar. "Context-Aware Data Prefetching in Mobile Service Environments". In *Proc. IEEE Fourth International Conference on Big Data and Cloud Computing (BdCloud)*, pages 214–221, 2014.

Chapter 3
Architecture of Data-Driven Personalized QoE Management

Abstract In this chapter, we propose a systematic architecture on data-driven personalized QoE management. A framework of the QoE management architecture is firstly introduced, which consists of two modules namely (1) training module and (2) control module. We also depict two models for the prediction of user preference, including Bayesian Graphic Model and Context Aware Matrix Factorization Model. A preliminary use case is deployed to demonstrate and evaluate the proposed architecture. Simulation results illustrate the superior performance of proposed architecture compared with traditional water-filling method.

3.1 Introduction

The ultimate target of QoE management, following the modeling, control and assessment of QoE, is to attain the control over QoE by optimization and controlling strategy. In order to maximize the quality of experience and the degree of end-user satisfaction, along with the expected utilization efficiency on limited resources of communication system, control strategy is expected to provide optimized services while controlling QoE dynamically and continuously. From the operators' perspective, the purpose of QoE management is both maintaining users satisfaction effectively and allocating available wireless network resources efficiently. It is always a challenging problem to optimize and manage QoE due to the non-negligible influence on QoE of many factors, including restricted network environment (limited network bandwidth, network variance, etc.), increasing data flow of mobile network, randomness of device location and service situation, and diversity of users profiles, requirement, and expectation. As a consequence, a comprehensive design from different perspectives is vital in the topic of QoE management.

Traditional QoE management and control only focused on the influence of resource such as network and QoS, but neglected the difference among the individual users. For example, for a man who is not Kobe' fans, a high-precision video and wide bandwidth allocated on Bryant motion video may not lead to high degree of satisfactory feelings because the user may simply not want to watch this video. The classic strategy of QoE management can not figure out this problem.

© The Author(s) 2017   21
Y. Wang et al., *QoE Management in Wireless Networks*, SpringerBriefs
in Electrical and Computer Engineering, DOI 10.1007/978-3-319-42454-5_3

With the development of mobile Internet, a huge amount of new smart devices and many service providers are emerging and growing rapidly. Win or lose heavily depends on the ways on the satisfaction of the various user demands. In particular with the advent of Web 2.0, we have entered into an unprecedented era of big data with explosively growing information. For example, YouTube has 2 billion video contents which grow at a speed of 60 h per minute meanwhile [1]. The big data leads to many new problems faced by video service providers. On the one hand, users may get confused or even feel bored to find the interested ones from vast amounts of videos, which affects user experience badly [2]. Furthermore, many new competitors enter the same business field to provide the similar services. It is meaningful for service providers to perceive the user's feeling, and provide the personalized service [3]. The characteristics of the personalized service includes: (1) The service is user-oriented. (2) The service is unique and varied along with the change of user requirements. (3) The service is diverse and discrepant [4].

Assessing QoE is one of key issues in the topic of QoE. The greatest difficulty to assess QoE from a user perspective is how to quantitatively use the user subjective information. In addition, a data-driven architecture for personalized QoE management is proposed in this chapter.

3.2   Framework of Data-Driven Personalized QoE Management

3.2.1   Basic Requirements

The prerequisites of the proposed architecture are listed in the following:

- A monitor to capture real-time information is contained. The real-time information includes the type of application in use and the status of QoS.
- A data mining scheme is maintained, which is capable to predict user preference/expectation with respect to the specific applications.
- Resource management is conducted according to the captured QoS status and the predicted preference/expectation information to maintain a reasonable degree of QoE.

Figure 3.1 illustrates the two components including the training module and the control module of the proposed architecture.

3.2.2   Training Module

The training module is designed to collect the subjective data, train and tune the user preference prediction model. The inputs of the model are obtained from a specific

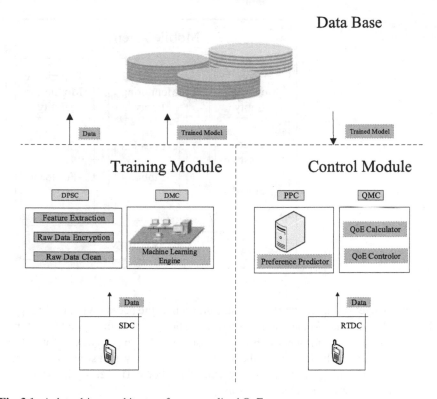

Fig. 3.1 A data-driven architecture for personalized QoE

user and a specific service, and the output is the user's expectation with respect to the service.

The training module is divided into three subcomponents described as follows.

(1) Subjective Data Collector (SDC)

SDC is responsible for collecting the data. The diversity of the data is the major concern when designing this subcomponent since the different types and dimensions of data should be integrated in the model training process. Due to end-user oriented characteristic of the data, the central part of such collector is a mobile agent designed to be physically deployed in the end-user device.

Such a mobile agent is logically divided into three entities: the QoS monitoring entity, the contextual monitoring entity and the experience monitoring entity. Besides, a data repository is created in the device to preserve the temporarily collected data, which is then uploaded to the cloud by synchronizing with the data processing and storage component. The framework of the mobile agent is presented in Fig. 3.2. The details of SDC are given as follows.

- The main function of the QoS monitoring entity is to measure the technical parameters, including (1) the device information, such as the operating system and

Fig. 3.2 Framework of
mobile agent [5]

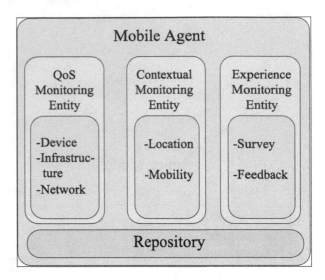

the screen size; (2) the network information, such as the access type, throughput,
delay, and jitter; and (3) the application information, such as the application type.

- The contextual monitoring entity is responsible for collecting context information
  of the application, which contains the location, the mobility, and some other available data from sensors on the device or around the body. This entity is extensible
  for a variety of newly developed sensors.
- The experience monitoring entity is in charge of the interaction with users by
  gathering explicit feedback of questionnaires, including both closed and open-
  ended question formats. The Experience Sampling Method (ESM) is chosen as the
  collection method in order to minimize the possible interference and the deviation.
  Moreover, with the further study of implicit experience measures, the experience
  probe can be extended with new modules and parameters, such as brain activity,
  heart activity, and eye movement.

(2) **Data Processing and Storing Component (DPSC)**

The data uploaded from SDC on mobile devices is preprocessed and stored in DPSC.
Once the new data is received, the damaged data is removed by an initial data cleaning
process. Then, a process of anonymization is done for pseudonyms and reduction of
location accuracy. And the further encryption is further proceeded for the reason of
the user privacy. The feature extraction is also completed in this component. And the
extracted features and encrypted data are preserved in a distributed database. Other
components in the architecture are only permitted to access the extracted features.

The user today generally has more than one mobile device and different login
accounts for distinct applications. To build connections among these individual
resources, a concept known as OpenID is established [6], in which a unique user
ID is allocated to a specific user uniquely as authenticated information. The user's

cell phone number is linked with other user specific information such as verifiable email address, etc. With unique user ID, DPSC merges data uploaded from each SDC according to a certain rule.

(3) Data Mining Component (DMC)

DMC is responsible for mining the user preference when they use certain services. The data-driven approach is used in this component to provide more effective prediction ability on user preference. Given a set of data, a user preference prediction model is trained by an offline mode. Most of previous QoE modeling approaches focused on how to model the mapping from network layer parameters to QoE parameters, but the role of the user preferences is ignored. Actually a specific user could prefer some services compared with others, which leads to the different satisfactory degrees for different services under the the same network layer condition.

In this section, we present several data mining models with regard to user preferences for DMC. A set of features is used for model training which is extracted in DPSC and reserved in a distributed database. Once the process of model training is finished, the model file is saved in a database and used by the control module for online user-preference prediction.

3.2.3 Control Module

The control module provides online service for monitoring and collecting both user and service information. The preferences for each user on different services are predicted by using the model derived in DMC. The control module is composed of three components namely (1) the real-time data collector, (2) the preference predictor and (3) the QoE controller.

(1) Real-Time Data Collector (RTDC)

Unlike SDC, the data collection is necessary to be finished in a very short period of time. Three types of data are gathered by RTDC including user ID, service in use and the network status. The software entity of RTDC is placed on the user mobile device, by which the user-service data are collected and sent to the Preference Prediction Component (PPC) and the network status information to the QoE controller.

(2) Preference Prediction Component (PPC)

The function of PPC is to provide prediction ability for user preference. The user-service data from RTDC is processed in PPC and user preference is predicted by applying prediction model trained in DMC. It should be noticed that there could be multiple models for handling different types of services. And the model selection is done automatically.

(3) QoE Management Component (QMC)

The predicted result of user preference by PPC is handed over to QMC. The status of the objective network is also identified by QMC. Given those information as the input, a QoE management model is derived which is summarized in Table 2.2. The QoE management is thus formulated as an optimization problem, and two types of optimization objective functions are selected including (1) maximizing the total QoE in the system, and (2) improving the QoE fairness for all users. Based on the results of optimization derivations, a QoE management scheme is designed by utilizing the status of the current system and the user preferences.

In summary, the whole process includes offline part and online part. Offline part is responsible to build up the user preference model by using data-driven approach. The SDC component distributed in each user's mobile device collects upload user specific information. And the DPSC component uses the data to train model. The function of online parts is provided by control module for online service. The whole process begins with monitoring of RTDC in the user device for the user action. Once the user begins to use a application/service, RTDC collects the realtime user/service information, which is then sent to PPC. PPC selects and uses the user preference model to predict the user preference. The resulting user preference is then utilized by QMC for QoE management on a system level.

Compared with the previous architecture, the main novelty of the proposed architecture is to involve a new concept, that is data driven user preference prediction. We present two kinds of modeling techniques for predictions.

3.3 Personalized Character Extraction: User-Service Preference

In this section, two models: Bayesian Graphic Model [7] and Context Aware Matrix Factorization Model [8] are introduced. Given the above discussion, the model is trained in the offline training stage, and used for online service within control module.

3.3.1 Bayesian Graphic Model (BGM)

The first modeling technique is Bayesian Graphic model, which is shown in Fig. 3.3, where subscripts i and j stand for user and service, respectively. This model contains three kinds of layers, which are, respectively, observable layer, hidden layer and prediction layer from the bottom to the top in turn. In the top layer, r_{ij} represents user x_i's real rating towards video y_j, $\hat{r}_{ij}$ represents user x_i's predicted rating towards video y_j. In the bottom layer, p_i and q_j represent user and video observed feature vectors, whose dimension is $p \times 1$ and $q \times 1$, respectively. In the middle layer, u_i and v_j represent user and video hidden (latent) feature vectors. The dimension of u_i

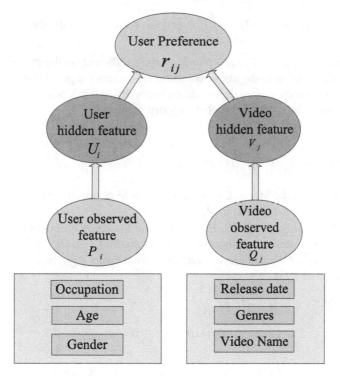

Fig. 3.3 User-service preference model [7]

and v_j are $r \times 1$. These three layers construct a Bayesian Graphic, in which r_{ij} is random variable which depends on the two random variables u_i and v_j. And u_i and v_j also depend on p_i and q_j, respectively. Thus this Bayesian Graphic model describes a probabilistic mapping from the observed features to the user's rating results. A Gaussian assumption is used in this model defined as follows.

$$r_{ij} \sim N\left(u_i^T v_j, \sigma^2\right),$$

$$u_i \sim MVN\left(Gp_i, \Sigma_u\right),$$

$$v_j \sim MVN\left(Dq_j, \Sigma_v\right),$$

where $N\left(\mu, \sigma^2\right)$ indicates the Gaussian distribution with expectation μ and variance σ^2; $MVN(U, \Sigma)$ indicates the multi-dimension Gaussian with expectation being U and covariance matrix being Σ; G and D indicate the coefficient matrixes with dimension being $r \times p$ and $r \times q$, respectively. Based on the above definitions, the probability relationship in BGM is as follows.

$$P\left(r_{ij}|p_i, q_j, \Theta\right) = \iint P\left(r_{ij}, u_i, v_j | p_i, q_j, \Theta\right) du_i dv_j, \qquad (3.1)$$

where $\Theta = (\sigma, G, \Sigma_u, D, \Sigma_v)$ represents all the coefficients in the model. The joint probability distribution in the above equation represents the generation model with the complete data set $\{r_{ij}, u_i, v_j\}$. According to conditional independence assumption of the hidden features, the joint probability distribution is

$$\begin{aligned} &P\left(r_{ij}, u_i, v_j | p_i, q_j, \Theta\right) \\ &= P\left(r_{ij} | u_i, v_j, p_i, q_j, \sigma\right) \cdot P\left(u_i | p_i, G, \Sigma_u\right) \cdot P\left(v_j | q_j, D, \Sigma_v\right). \end{aligned} \qquad (3.2)$$

In the training step, the Monto Carlo Expectation-Maximization (EM) algorithm is adopted in this case to train the model.

Denote the training set as $\{R^T, P^T, Q^T\}$. After the above training process, coefficients Θ and the posterior distribution of latent variables $P(U|R^T, P^T, Q^T, \Theta)$ and $P(V|R^T, P^T, Q^T, \Theta)$ are achieved. The prediction target is $\hat{r}_{ij}$, based on the BGM, the distribution of $\hat{r}_{ij}$ is

$$\begin{aligned} &\mathrm{P}\left(\hat{r}_{ij}|R^T, P^T, Q^T, \Theta\right) \\ &= \iint P(\hat{r}_{ij}, U, V | R^T, P^T, Q^T, \Theta) dU dV \\ &= \iint P\left(\hat{r}_{ij}|U, V, \Theta\right) \mathrm{P}\left(U|R^T, P^T, Q^T, \Theta\right) \\ &\quad \mathrm{P}\left(V|R^T, P^T, Q^T, \Theta\right) dU dV, \end{aligned} \qquad (3.3)$$

with this distribution, the $\hat{r}_{ij}$ can be predicted.

Integrations w.r.t U and V lead to a very high computation complexity. So, some approximate methods are preferred. The expectation of hidden variable $\bar{U}$ and $\bar{V}$ can be estimated by using their posterior distribution. Then the probability density function of $\hat{r}_{ij}$ is approximated by

$$P\left(\hat{r}_{ij}|R^T, P^T, Q^T, \Theta\right) \approx \mathcal{N}\left(\hat{r}_{ij}|\bar{u}_i\bar{v}_j, \sigma\right). \qquad (3.4)$$

So the approximation of $\hat{r}_{ij}$ is $\bar{u}_i\bar{v}_j$.

Once the BGM is used online, x_i or y_j represents the new user or a new video, respectively. We use observed features p_i and q_j to estimate u_i, v_j and Θ. And the cold start problem can be solved.

3.3.2 Context Aware Matrix Factorization Model

Before the discussion on context aware matrix factorization model, we present a information gain based factor pre-selection process. Given a set of factors, it is meaningful to proceed a factor pre-selection step to filter out some less influential factors, which makes the modeling work more easier. The different applications/services have their own requirements. There are actually no common set of factors suitable for all applications/services. In this section, we use a information gain to select contextual factors. Given a contextual factor variable denoted by $D = \{d_k, 1 \leq k \leq K\}$ and $R = \{r_i, 1 \leq i \leq I\}$, and the information gain is formulated as follows.

$$
\begin{aligned}
Gain(R, D) &= H(R) - H(R|D) \\
&= H(R) - \sum_k p(D = d_k) H(R|D = d_k)
\end{aligned}
\tag{3.5}
$$

where $Gain(R, D)$ indicates the information gain between the rating R and the contextual factor D; $H(R)$ indicates the information entropy of the rating R; $H(R|D)$ indicates the conditional entropy of the rating R given the contextual factor D; $H(R|D = d_k)$ indicates the conditional entropy of the rating R when the contextual factor $D = d_k$.

$H(R)$ depicts the uncertainty of rating R, meantime $H(R|D)$ depicts the uncertainty of rating R given the contextual factor D. Naturally, $Gain(R, D)$ represents the decreased uncertainty of rating R given the contextual factor D. The higher $Gain(R, D)$ means the stronger influence on predicting rating R determined by the contextual factor D. Consequently, the contextual factors with lower information gain are regarded as less influential factors, which will be removed from the set of factors.

The matrix factorization (MF) is formulated as:

$$
\hat{r}_{ij} = p_i^T q_j
\tag{3.6}
$$

where $\hat{r}_{ij}$ indicates the predicting rating of user i to item j; p_i denotes features of user i; q_j indicates features of item j.

And the context aware matrix factorization model extends MF model to integrate the context information. The context aware MF model is used for user preference prediction under different contexts, which is shown in Fig. 3.4. Subscripts i and j are used to represent user and service, respectively.

Here, we simply divide the contextual factors into three types: the user context, the item context and the interaction context. The detailed contextual factors are shown in the following by using movie recommendation as example:

- User Context: represents user attributes (Age, Gender, Occupation, Faith, Country, etc.). Generally, they are acquired by users' registration information.
- Item Context: represents movie attributes (Budget, Director, Genres, Actors, Language, etc.). Generally, they are acquired by publishers.
- Interaction Context: represents the condition of users or environment when watching movies (Time, Mood, Location, Day type, Social, etc.). Generally, they are acquired by user behavior analysis and server records.

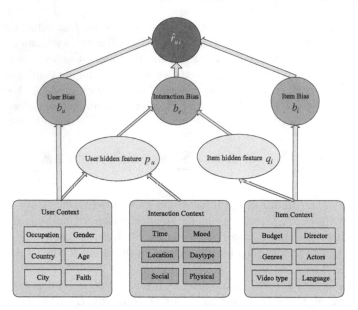

Fig. 3.4 The structure of context aware matrix factorization model [8]

Using the definitions defined in the last section, we further define $\hat{r}_{ij}$ as

$$\hat{r}_{ij} = \mu + b_i + b_j + b_e = \\ \mu + b_i(c_i) + b_j(c_j) + [p_i(c_i, c_e)]^T q_j(c_j)$$ (3.7)

as the predicted preference of user i to application/service j. And the relevance between predicted rating and each context is simplified as the sum of four parts: global average, user bias, item bias, and user-item interaction bias. μ is the average score of all known ratings. b_i and b_j indicate the observed deviations of user i and item j, respectively. p_i and q_j are two K-dimensional vectors that represent the features of user i and item j, $p_i \in \mathbb{R}^{K \times 1}$, $q_j \in \mathbb{R}^{K \times 1}$.

Given the definition of the predicting rating $\hat{r}_{ij}$, subsequently the training method is shown. The method conducts a loss function related to the parameters in $\hat{r}_{ij}$, which can be minimized using stochastic gradient descent method to train the model, more details of this method are shown in [8]. Consequently, parameters of $\hat{r}_{ij}$ above are solved and then the predicting rating can be calculated using Eq. 3.7 along with new data.

3.4   Personalized QoE Model and Example User Case

In order to validate the proposed architecture [5], we carry out a preliminary use case. In this case, the service scenario is online video. The data includes the user information collected from some volunteers from the university campus and the video information. The information data of users includes gender, age, occupation, and preferred video type. The videos are categorized into 16 classes and the video information includes the released dates. The context features we considered include whether the user is indoor or outdoor, whether the user is walking or sitting, and whether the user watches video using a smartphone or a tablet.

In this case, a two-step QoE modeling method is presented, which depends not only on network layer parameters, but also on user-service preferences. In the first step, the user-service preference is modeled and the preference value is predicted as the feature of QoE modeling. In the second step, QoE is modeled according to both network layer parameters and user preferences. A sigmoid function is used to formulate the mapping from network layer parameters and user preferences to QoE as follows:

$$QoE = \frac{\theta}{1 + e^{(-\alpha S + \beta r_{ij} + \gamma)}},\qquad(3.8)$$

where θ, α, β and γ are a set of parameters to control the relationship between input and output of sigmoid function.

Without loss of generality, we simply select bit rate to represent QoS. We assume that the users are randomly distributed in the coverage areas of 5 APs and each AP has an upper bound of the total bit rate. An optimization problem is formulated to describe the maximization of the total QoE. By solving the problem, a properly tuned bit rate could be achieved for the optimization of QoE.

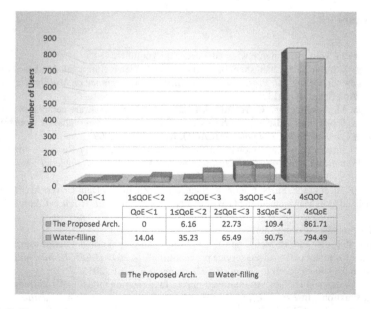

	QoE<1	1≤QoE<2	2≤QoE<3	3≤QoE<4	4≤QoE
The Proposed Arch.	0	6.16	22.73	109.4	861.71
Water-filling	14.04	35.23	65.49	90.75	794.49

Fig. 3.5 QoE comparison

We compare the QoE distribution in the proposed architecture with that of the traditional water-filling algorithm. In traditional water-filling algorithm, iterative bit rate allocation solution is obtained as a result of total QoE maximization without consideration of the user expectation/preference, which results in more bit rate allocated to the user with lower QoE. Simulation results show improved performance of the system compared with the traditional resource allocation method, with an increase of the total QoE by 20 and 96% of the users obtaining a higher QoE, as shown in Fig. 3.5.

3.5 Summary

In this chapter, we propose a systematic data-driven personalized QoE management architecture for QoE optimization. A framework of the QoE management architecture is first introduced, which consists of a training module and a control module. We also employ two models to predict user preference, including Bayesian Graphic Model and Context Aware Matrix Factorization Model. To evaluate the proposed architecture, we deploy a preliminary use case. Simulation results illustrate that the performance of the proposed architecture increases by 20% in the total QoE, while 96% of the users obtain a higher QoE compared with the traditional water-filling method.

References

1. Encyclopedia. Youtube. http://en.wikipedia.org/wiki/youtube/.
2. D. Zhengyu, S. Jitao, and Changsheng X. "Personalized video recommendation based on cross-platform user modeling". In *Proc. of IEEE International Conference on Multimedia and Expo (ICME)*, America, July. 2013.
3. S. Kisoon, M. Aekyung, and C. Yongil. "A case study of context-aware personalized services on IMPACT platform: Adsonmap services". In *Proc. of IEEE International Conference on Advanced Communication Technology (ICACT)*, Ireland, Feb. 2009.
4. Z. Hua. "Construction of a Personalized Service Oriented Learning Resource Management System Framework". In *Proc. of IEEE International Workshop on Knowledge Discovery and Data Mining (WKDD)*, Saudi Arabia, Jan. 2008.
5. Y. Wang, P. Li, J. Lei, Z. Su, N. Cheng, X. Shen, and P. Zhang. "A Data-Driven Architecture for Personalized QoE Management in 5G Wireless Networks". In *IEEE Wireless Communications*. IEEE.
6. OpenID. http://www.openid.net.
7. Y. Wang, P. Li, H. Tao, R. Meng, and J. Liu. "Bayesian graphic model based user preference prediction for future personalized service provisioning". In *2015 IEEE/CIC International Conference on Communications in China (ICCC)*, pages 1–6. IEEE, 2015.
8. J. Liu, Y. Wang, and H. Tao. "An improved matrix factorization model under multidimensional context situation". In *2015 IEEE/CIC International Conference on Communications in China (ICCC)*, pages 1–6. IEEE, 2015.

Chapter 4
QoE-Oriented Resource Allocation in Wireless Networks

Abstract User-subjective experience and personalized preference play a very important role in QoE. In this chapter, we intend to develop personalized QoE management by considering the user preference. The proposed personalized QoE model integrates objective QoS parameters and subjective user preference to evaluate user QoE and thus a more reasonable QoE assessment is achieved. Then, the personalized QoE model is applied to the sequential resource allocation and a more specialized and refined management scheme can be attained. For both conventional and personalized resource allocation schemes, numerical simulations are conducted to present the effectiveness of the proposed algorithms.

4.1 Background

Radio resource allocation or scheduling algorithms are in charge of deciding when and how to assign limited radio resources to network users, in order to maximize network capacity, or to ensure QoS requirements, user fairness, etc. Traditional basic radio resource scheduling algorithms, such as Round Robin scheduling (RR) algorithm, Maximum Throughput (MT) algorithm and Proportional Fair (PF) algorithm [1–4] are not suitable for multimedia services any more due to the ignorance of service quality. Then, QoS is proposed as an indicator to describe objective service quality, including Assignment Reservation Priority (ARP), Guaranteed Bit Rate (GBR) and QoS Class Identifier (QCI) in the 3GPP standard [5], where QCI is a pointer pointing to a more detailed QoS parameter table, including service priority, budget of L2 packet delay, and packet loss rate of L2 packet [2]. However, in the context of increasing competition among network operators and service providers, it is important to provide users satisfied QoE in order to keep and attract more users. QoE has become a crucial indicator to evaluate comprehensive performance of wireless communication systems. Hence, the focus of radio resource management has been shifted from improving objective system performance to promoting user subjective experience.

To illustrate the research status of radio resource management, we mainly divide them into three types according to scheduling objectives: QoS-based, QoE-based,

© The Author(s) 2017
Y. Wang et al., *QoE Management in Wireless Networks*, SpringerBriefs
in Electrical and Computer Engineering, DOI 10.1007/978-3-319-42454-5_4

and energy efficiency based. More details of these resource management schemes are given as follows.

4.1.1   QoS-Based Radio Resource Management Strategies

Real-time multimedia applications (such as voice, web browsing, and file download-ing services) have become the most important services in wireless communication systems, motivating QoS parameters to be considered in wireless resource manage-ment strategies. For QoS performance indicators (like transmission rate, network latency, packet loss rate, etc.), researchers have proposed a number of radio resource scheduling strategies for different optimization objectives and applicable scenar-ios. References [6–10] proposed QoS-based resource allocation strategies in order to fulfill user requests such as upper delay time and packet loss rate. In particular, a detailed and comprehensive analysis about QoS-based resource scheduling policies was given in [6]. The EXP and LOG rules were proposed and claimed to have a great prospect in the delay-sensitive OFDM system [6].

4.1.2   QoE-Based Radio Resource Management Strategies

For multi-media services with strong content relevance, QoS can not reflect user subjective experiences accurately due to the ignorance of user factors, motivating the research focus to shift from QoS to QoE. In [11], based on human stimulated perception model in the psychology field, a QoE model in the form of log func-tion was proposed, mapping certain QoS parameters (such as data rate, delay and other parameters) into the MOS (mean opinion score) value by log function. Most existing QoE-based radio resource management strategies follow such QoS to QoE mapping model currently, turning data rate throughput maximization problem into MOS maximization problem. Reference [12] studied a MOS-based radio resource management strategy based on game theory in OFDMA systems. For QoE optimiza-tion, bitrate-MOS mapping models were adopted to match data rate to QoE for video, voice, and data service, respectively. In [13], a radio resource management strategy in MIMO-OFDM system for data transmission service was proposed based on QoE mapping model.

The process to investigate a QoE-based resource resource management can be summarized as follow: (1) Determine the QoE estimation method and develop the system model. (2) Formulate the resource allocation problem according to specific objectives decided by network managers. (3) Design a resource allocation strategy to obtain the optimal or suboptimal solution of the formulated resource allocation problem.

4.1.3 Energy Efficiency-Based Radio Resource Management Strategies

With the rapid development of mobile communication, the global warming issues caused by carbon dioxide emission increasing have drawn much attention for future mobile communication systems. It is challenging to design resource management schemes which can provide good service performance and reduce energy consumption simultaneously. Therefore, energy efficiency defined as the ratio of system throughput or other performance indexes to whole energy consumption turns into a hot topic in wireless communications.

There have been a lot of radio resource management strategies designed in physical layer and MAC layer or cross layer in order to improve system energy efficiency. Reference [14] first introduced the theoretical limits of energy efficiency according to information theory, and analyzed energy efficient design of Multiple-Input Multiple-Output (MIMO) and relay system. Moreover, radio resource allocation strategies based on signal and data symbols were further discussed in that paper. In [15], four kinds of trade-off of basic performance indexes, including deployment efficiency versus energy efficiency, spectrum efficiency versus energy efficiency, bandwidth versus energy and latency versus energy are studied. As for radio resource scheduling based on energy efficiency, [16] studied an energy efficient resource allocation strategy in uplink and downlink OFDMA system. To optimize overall energy efficiency in the downlink and maximize the lowest energy efficiency of users in the uplink with QoS constraints, an optimal algorithm and a suboptimal low complexity algorithm were proposed based on the property of energy efficiency problem.

A radio resource scheduling algorithm with high energy efficiency in multi-cell OFDMA downlink system under the BS collaboration was proposed in [17]. In particular, on the basis of basic energy efficiency problem, limited backhaul bandwidth, minimum user data rate requirement and other constraints were added. To solve the problem, it first transformed the fractional form optimization problem into an equivalent subtraction optimization problem, and then a low complexity iterative algorithm was proposed. Conclusions were drawn according to simulation results that high spectral efficiency did not necessarily guarantee high energy efficiency and that energy efficiency can be improved by increasing the backhaul bandwidth to a certain extent. A joint optimization problem of QoS and energy efficiency in downlink OFDMA network was studied in [18]. QoS parameters such as delay were transformed into equivalent transmission rate in order to formulate the energy efficiency optimization problem. The upper bound of the original problem was derived and then suboptimal solution was obtained to realize the tradeoff between energy efficiency and delay.

4.2   Traditional QoE-Based Resource Allocation Mechanism

4.2.1   QoE Metric Model

Three typical applications are considered in this chapter including audio stream, data stream and video stream. Each user is assumed to experience a single application service. The utility function of QoE is expressed in terms of MOS estimated for each stream. The relationship between MOS and user satisfaction is illustrated in Fig. 4.1.

As an example, details of estimating MOS for three applications are described below.

4.2.1.1   Audio Stream

Different methods have been proposed to evaluate user experience of audio stream, such as Perceptual Evaluation of Speech Quality (PESQ) and Pseudo-Subjective Quality Assessment (PSQA). We adopt the PESQ-based MOS estimation method proposed in [20], formulating MOS as a function of the transmission rate R and the packet error probability (PEP). Figure 4.2 shows experimental curves of MOS versus PEP for different voice codecs.

Accounting for the high data rate of LTE, a reasonable assumption is that all the audio stream are encoded by G.711. Thus the MOS function of PEP can be formulated as

$$MOS = 0.5 * log(\frac{1}{1 + 60 * PEP}) + 4.3. \tag{4.1}$$

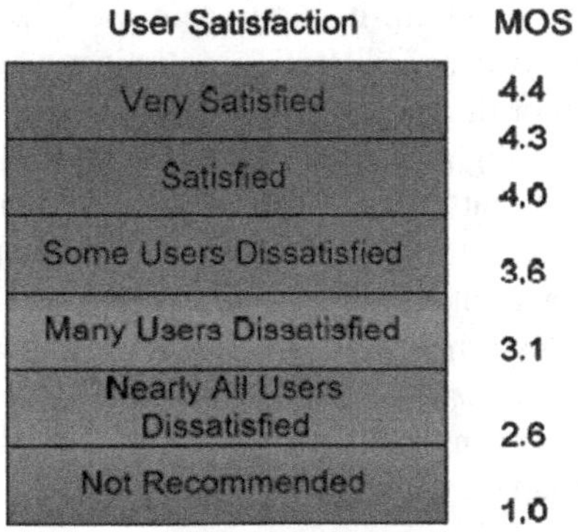

Fig. 4.1 Relationship between MOS and user satisfication [19]

User Satisfaction	MOS
Very Satisfied	4.4
	4.3
Satisfied	4.0
Some Users Dissatisfied	3.6
Many Users Dissatisfied	3.1
Nearly All Users Dissatisfied	2.6
Not Recommended	1.0

Fig. 4.2 PESQ-based MOS versus packet error probability (PEP) for different transmission rate [20]

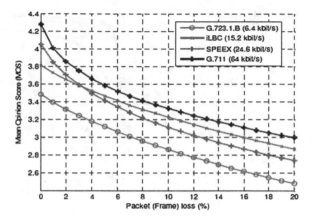

4.2.1.2 Data Stream

Surfing webs and downloading files on the Internet both belong to data stream service. For this type of service, the waiting time until the download process is finished is the most important influencing factor for user's QoE. Download bandwidth, user's throughput here, has direct relationship with the waiting time for the data downloading. It is concluded in [11] that MOS increases logarithmically with increasing data throughput. In addition, user's tolerance for waiting time is influenced by the size of the file. The MOS function is modeled in [11] as follows.

$$MOS = \frac{0.775}{\sqrt{s}} ln(T) + 1.268,$$ (4.2)

where, s stands for the file size with unit of MB and T is the throughput of a user.

4.2.1.3 Video Stream

Video stream service has high requirement for data rate, continuity and real-timing. MOS for video service is complicated to evaluate because users' QoE is related to not only transmission rate but also the content of the video. Reference [21] points out that Video Structural Similarity Index (VSSIM) can evaluate users QoE more reasonably than images Mean Square Error (MSE) or Peak Signal Noise ratio (PSNR). That is because human eyes are more sensitive to the distortion of images structure than that of images pixels. According to the data from Video Quality Expert Group (VQEG) [22], MOS can be obtained according to VSSIM as follows.

$$MOS = \begin{cases} 1 & VSSIM < 0.7 \\ 12.5 \cdot VSSIM - 7.75 & 0.7 \leq VSSIM \leq 0.98 \\ 4.5 & VSSIM > 0.98. \end{cases}$$ (4.3)

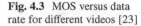

Fig. 4.3 MOS versus data
rate for different videos [23]

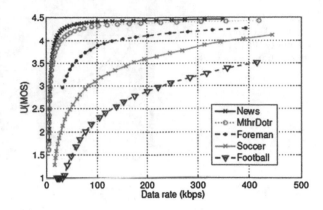

Figure 4.3 shows the video utility curve for five different video sequences. It can be concluded that videos with higher dynamic scenes are more demanding for data rate than static videos.

4.2.2   System Model

In a multi-service single cellular downlink OFDM system with a BS communicating with K users sharing N subcarriers. The total bandwidth of the system W is equally divided into N subcarriers, with the bandwidth of each subcarrier $B = W/N$. In order to avoid co-channel interference between different users in the same cell, each subcarrier in a sub-frame can be assigned to only one user. However, one user can be assigned more than one subcarrier.

Denote $p_{i,n}$ and $G_{i,n}$ as the instantaneous transmit power and channel gain of user i on subcarrier n, respectively. $G_{i,n}$ represents the sum effect of path loss, shadow fading and small-scale fading. $I_{i,n}$ is the inter-cell interference. According to Shannon's formula, the maximum available data rate for user i on subcarrier n is

$$r_{i,n} = B\log_2(1 + \frac{p_{i,n} \cdot G_{i,n}}{N_0 B + I_{i,n}}), \qquad (4.4)$$

where N_0 is the white Gaussian noise power spectrum density. Hence, the total data rate of user i can be given by

$$R_i = \sum_{n=1}^{N} a_{i,n} \cdot r_{i,n}, \qquad (4.5)$$

where $a_{i,n}$ indicates whether subcarrier n is allocated to user i. $a_{i,n} = 1$ if the subcarrier n is scheduled to user i, otherwise $a_{i,n} = 0$. Moreover, each subcarrier can be allocated to at most one user $\sum_{i=1}^{K} a_{i,n} = 1$.

4.2.3 Problem Formulation

In order to achieve system QoE maximization, the resource allocation problem can be formulated as follows:

$$\max_{a_{i,n}, P_{i,n}} : \sum_{i=1}^{K} MOS_i,$$

$$C1 : \sum_{i=1}^{K} P_i \leq P_{total},$$

$$C2 : 0 \leq P_i \leq P_{max}, \forall i \in \{1, 2...K\},$$

$$C3 : a_{i,n} \in \{0, 1\}, \forall i \in \{1, 2...K\}, \forall n \in \{1, 2...N\},$$

$$C4 : \sum_{i=1}^{K} a_{i,n} = 1, \forall n \in \{1, 2...N\}.$$

(4.6)

4.2.4 Resource Allocation Strategy

Weber–Fechner Law (WFL) [24] illustrates a logarithmic relationship between user QoE and data throughput, which implies that a marginal effect does exist for QoE and transmission power. Meanwhile, the relationship between throughput and transmission power is also formulated as a logarithmic function. Therefore, MOS will exhibit a strong diminishing marginal benefit in terms of power. More specifically, there is limited MOS increase for a user whose MOS being already high with a certain amount improvement of power. While a low MOS will have a stronger enhancement with the same amount of power improvement. Hence, a QoE-based resource allocation algorithm with low complexity is presented in Fig. 4.4.

The $M+$ in the flowchart represents the achieved MOS gain if power allocation increases by ΔP, which can be given by $\Delta M_i^+ = MOS_i^m(P_i^m + \Delta P) - MOS_i^m(P_i^m)$ and $M-$ stands for the MOS degradation if power allocation decreases by ΔP.

4.2.5 Simulation and Analysis

To validate the effectiveness of the QoE-based resource allocation strategy, a simulation experiment is conducted and the Round-Robin strategy is adopted as the baseline

Fig. 4.4 QoE-based
resource allocation strategy

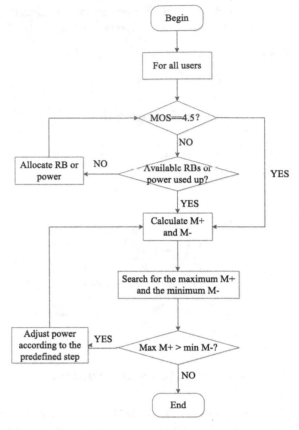

Table 4.1 Experiment
parameters

Carrier frequency	2 GHZ
Fast fading	Rayleigh fading
User number	10
Subcarrier number	200
Subcarrier bandwidth	10 kHZ
Noise power spectrum density (N0)	10^{-13} W/HZ

strategy for comparison. Simulation parameters are listed in Table. 4.1. Figure 4.5
presents the Cumulative Distribution Function (CDF) of user QoE. It can be concluded from Fig. 4.5 that the proposed QoE-based resource allocation strategy can
acquire better user QoE than QoE-ignorant baseline strategy.

Fig. 4.5 Cumulative
Distribution Function of user
QoE

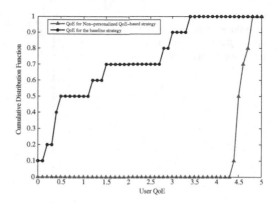

4.3 Personalized QoE-Based Resource Allocation Mechanism

To conduct personalized QoE-based management, a personalized QoE model is established first. The personalized QoE model proposed and validated in Chap. 3 is used here to instruct personalized QoE-based resource allocation. For personalized QoE management, a comprehensive consideration of both objective and subjective factors is used to evaluate user QoE. User preference is used to quantize subjective factors and QoS is used to quantize the objective factors. The QoE model is developed using the popular sigmoid function with respect to user's preference r_{ij} and QoS denoted by S, i.e., as is shown below.

$$QoE = \frac{\theta}{1 + e^{(-\alpha S + \beta r_{ij} + \gamma)}},\qquad(4.7)$$

where α, β, γ, and θ are the parameters constraining the quantization of QoE and they can be learned by implementing experiments and data analysis.

For sequential personalized QoE-based resource allocation purpose, the personalized QoE model is utilized to replace the QoE formulation in optimization problem (4.6), where user preference data derives from the user preference prediction model and bit rate is utilized to represent QoS parameters. A simulation is employed to depict the results comparison of non-personalized QoE-based strategy and personalized QoE strategy. Final user actual QoE distribution and data throughput distribution and their average values are shown in Figs. 4.8 and 4.9. And Figs. 4.6 and 4.7 depict the Cumulative Distribution Function of user actual QoE and data throughput, respectively. Obviously, more refined and precise QoE management is attained for personalized strategy because resources can be utilized more reasonably and efficiently by considering user personality and more system benefit can be reached accordingly.

Fig. 4.6 Cumulative
Distribution Function of user
actual QoE for personalized
and non-personalized
strategy

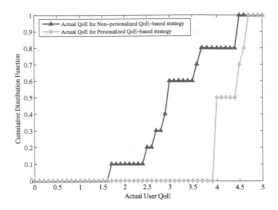

Fig. 4.7 Cumulative
Distribution Function of user
data throughput for
personalized and
non-personalized strategy

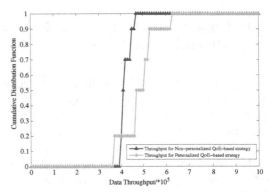

Fig. 4.8 User actual QoE
for personalized and
non-personalized strategy

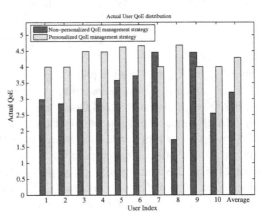

Fig. 4.9 User data throughput for personalized and non-personalized strategy

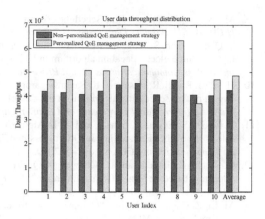

4.4 Summary

In this chapter, a brief introduction to radio resource allocation research status along with the developing direction is given first. Then, QoE-based resource allocation problem is investigated from both non-personalized and personalized aspects. A QoE-based resource allocation algorithm is proposed to attain suboptimal solution for QoE maximization problem. Simulation results validate the proposed algorithm by comparing QoE-based strategy with round robin strategy. Furthermore, a comparison of personalized and non-personalized QoE-based strategy is given, indicating that a more detailed and refined management is accomplished by considering user preference.

References

1. P. Kela, J. Puttonen, N. Kolehmainen, T. Ristaniemi, T. Henttonen, and M. Moisio. "Dynamic packet scheduling performance in UTRA long term evolution downlink". In *Proc. IEEE 3rd International Symposium on Wireless Pervasive Computing*, pages 308–313, 2008.
2. G. Monghal, K. I. Pedersen, PE Mogensen, et al. "QoS oriented time and frequency domain packet schedulers for the UTRAN long term evolution". In *Proc. IEEE Vehicular Technology Conference*, pages 2532–2536, 2008.
3. Y. Lin and G. Yue. "Channel-adapted and buffer-aware packet scheduling in LTE wireless communication system". In *Proc. IEEE 4th International Conference on Wireless Communications, Networking and Mobile Computing*, pages 1–4, 2008.
4. R. Kwan, C. Leung, and J. Zhang. "Proportional fair multiuser scheduling in LTE". *IEEE Signal Processing Letters*, 16(6):461–464, 2009.
5. D ETSI. "General Packet Radio Service (GPRS), service describtion; stage 2", 1998.
6. B. Sadiq, S. J. Baek, and G. De Veciana. "Delay-optimal opportunistic scheduling and approximations: The log rule". *IEEE/ACM Transactions on Networking (TON)*, 19(2):405–418, 2011.
7. J. Park, S. Hwang, and H. Cho. "A packet scheduling scheme to support real-time traffic in OFDMA systems". In *Proc. IEEE 65th Vehicular Technology Conference*, pages 2766–2770, 2007.

8. M. Assaad. "Frequency-Time Scheduling for streaming services in OFDMA systems". In *Proc. IEEE 1st IFIP Wireless Days*, pages 1–5, 2008.
9. Y. Qian, C. Ren, S. Tang, and M. Chen. "Multi-service QoS guaranteed based downlink cross-layer resource block allocation algorithm in LTE systems". In *Proc. International Conference on Wireless Communications & Signal Processing*, pages 1–4, 2009.
10. H. A. M. Ramli, R. Basukala, K. Sandrasegaran, and R. Patachaianand. "Performance of well known packet scheduling algorithms in the downlink 3GPP LTE system". In *Proc. IEEE 9th Malaysia International Conference on Communications (MICC)*, pages 815–820, 2009.
11. P. Reichl, S. Egger, R. Schatz, and A. D'Alconzo. "The logarithmic nature of QoE and the role of the Weber-Fechner law in QoE assessment". In *Proc. IEEE International Conference on Communications (ICC)*, pages 1–5, 2010.
12. C. Sacchi, F. Granelli, and C. Schlegel. "A QoE-oriented strategy for OFDMA radio resource allocation based on Min-MOS maximization". *IEEE Communications Letters*, 15(5):494–496, 2011.
13. M. Li, Z. Chen, and Y. Tan. "Qoe-aware resource allocation for scalable video transmission over multiuser MIMO-OFDM systems". In *Proc. IEEE Visual Communications and Image Processing (VCIP)*, pages 1–6, 2012.
14. G. Y. Li, Z. Xu, C. Xiong, C. Yang, S. Zhang, Y. Chen, and S. Xu. "Energy-efficient wireless communications: tutorial, survey, and open issues". *IEEE Wireless Communications*, 18(6):28–35, 2011.
15. Y. Chen, S. Zhang, S. Xu, and G. Y. Li. "Fundamental trade-offs on green wireless networks". *IEEE Communications Magazine*, 49(6):30–37, 2011.
16. C. Xiong, G. Y. Li, S. Zhang, Y. Chen, and S. Xu. "Energy-efficient resource allocation in OFDMA networks". *IEEE Transactions on Communications*, 60(12):3767–3778, 2012.
17. D. W. K. Ng, E. S. Lo, and R. Schober. "Energy-efficient resource allocation in multi-cell OFDMA systems with limited backhaul capacity". *IEEE Transactions on Wireless Communications*, 11(10):3618–3631, 2012.
18. C. Xiong, G. Y. Li, Y. Liu, and S. Xu. "QoS driven energy-efficient design for downlink OFDMA networks". In *Proc. IEEE Global Communications Conference (GLOBECOM)*, pages 4320–4325, 2012.
19. ITUT Rec. "G. 107-The E Model, a computational model for use in transmission planning". *International Telecommunication Union*, 8:20–21, 2003.
20. S. Khan, S. Duhovnikov, E. Steinbach, M. Sgroi, and W. Kellerer. "Application-driven cross-layer optimization for mobile multimedia communication using a common application layer quality metric". In *Proc. the 2006 international conference on Wireless communications and mobile computing*, pages 213–218. ACM, 2006.
21. M. Shehada, S. Thakolsri, Z. Despotovic, and W. Kellerer. "QoE-based cross-layer optimization for video delivery in long term evolution mobile networks". In *Proc. 14th International Symposium on Wireless Personal Multimedia Communications (WPMC)*, pages 1–5, 2011.
22. A. M. Rohaly, J. Libert, P. Corriveau, A. Webster, et al. "Final report from the video quality experts group on the validation of objective models of video quality assessment". *ITU-T Standards Contribution COM*, pages 9–80, 2000.
23. S. Thakolsri, S. Cokbulan, D. Jurca, Z. Despotovic, and W. Kellerer. "QoE-driven cross-layer optimization in wireless networks addressing system efficiency and utility fairness". In *Proc. 2011 IEEE GLOBECOM Workshops (GC Wkshps)*, pages 12–17, 2011.
24. P. Reichl, S. Egger, R. Schatz, and A. D'Alconzo. "The logarithmic nature of QoE and the role of the Weber-Fechner law in QoE assessment". In *Proceedings of IEEE International Conference on Communications (ICC)*, pages 1–5, 2010.

Chapter 5
Implementation and Demonstration of QoE Measurement Platform

Abstract This chapter describes the state-of-the-art of QoE experiment. Then using a streaming media application as an example, how to design subjective experiment is presented in particular to the QoE-related factors and measurement criterion definition. The detailed procedure and infrastructure is then presented and illustrated. Finally, a conclusion is given.

5.1 Introduction

Although many research works are involved in QoE, it is still an open problem on how to measure QoE quality. Since the user-subjective information such as user preference is highly related with QoE quality, how to balance the influences between objective and subjective factors is a key issue.

5.2 Related Work

According to the wireless network environment which QoE measurement work depends on, the measurement is divided into three categories, namely, commercial network, laboratory network, and simulation network.

5.2.1 Measurement Under Commercial Network Environment

The data from commercial network is of course realistic data. Because data is usually user-dependent. The relevant privacy regulations should be respected. In addition, the end-users obviously do not want to be disturbed when they are using the applications. Some smart ways should be found to promote the end-users to response their subjective feelings. Schuurman et al. discusses a concept of "living laboratory,"

© The Author(s) 2017
Y. Wang et al., *QoE Management in Wireless Networks*, SpringerBriefs
in Electrical and Computer Engineering, DOI 10.1007/978-3-319-42454-5_5

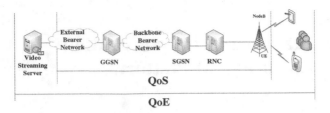

Fig. 5.1 QoE and QoS for video streaming service in 3G networks [5]

in which they create a real-life community within a commercial market [1]. With a large amount of users involved, the measurement result becomes more accurate. De Moor et al. has proposed a distributed architecture to monitor the QoS, the context information and the subjective user experience based on the functional requirements which is related to real-time experience measurements in real-life settings [2–4]. This approach aims to evaluate all relevant QoE-dimensions in a mobile context. X. Yu et al. has introduced an end-to-end, no-reference, and real-time QoE prediction model for video streaming in 3G networks [5–7]. In this model, comprehensive parameters from the network layer, the application layer, decoded videos and the user equipments are collected and integrated. This model has a good performance on the simulation results in terms of accurate prediction mean opinion score (MOS), small root mean squared error (RMSE), and low time-consuming. Figure 5.1 describes QoS and QoE for video streaming service in 3G networks. However, due to the difficulties in collecting operators' actual network data, there is a big gap between the data that we have actually got and that we ideally need. How to obtain more comprehensive and effective data that we really need is also a tough challenge.

5.2.2   Measurement Under Laboratory Network Environment

Due to the problems discussed in the last section, many researchers incline to establish the laboratory network environment based on their requirement. With a laboratory network environment, the researchers have a better control on the whole infrastructure to facilitate QoE measurement. In Ref. [8], an active feedback measurement scenario is built up in a laboratory network which contains the terminal domain and the server domain. When the experiment is in progress, users will shake their mobile phones if they are dissatisfied with the quality of the videos. Then, the feedback information will be sent back via the terminal. By this way, the feedback information is collected and the value of the QoE parameter is calculated for each user. In Ref. [9], another measurement scenario is also constructed for short-form videos considering the memory effect. Plenty of subjective tests is proceeded within the laboratory network environment. Figure 5.2 shows the whole architecture. It is observed that the primacy factor has more influence than recency factor on short-form videos.

Fig. 5.2 Laboratory network

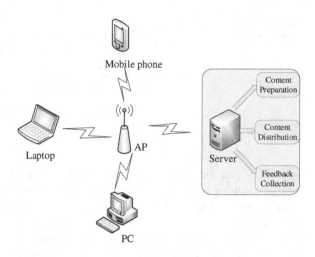

In Ref. [10], the QoE unfairness issue is studied by investigating how the segment duration of the video content affects the MPEG-DASH user QoE level. Amounts of experiments are conducted by using different segment duration on an interactive DVB-T testbed. The experimental results show a strong correlation between the video segment duration and the MPEG-DASH user QoE level. That is, a longer segment duration is preferred to achieve a higher QoE level than a shorter one. In Ref. [11], a VMOS model is proposed to predict the video streaming quality. An accurate end-user subjective perception evaluation is claimed on video streaming with low residual error. An adaptive laboratory test environment is also set up for the subjective measurement. And the result indicates that the correlation between the VMOS score and MOS is no less than 0.9. Thus, the VMOS score is a pretty good metric to evaluate end-user perception. In Ref. [12], QoE measurement is focused on the influence of test duration investigated on user fatigue and the reliability of user ratings. Three typical QoE lab studies are conducted with different task profiles including audio, video and web task. The different network conditions are evaluated. Given the results, long term active testing does not influence participants' quality gradings even with the presence of measurable signs of fatigue. Thus, for comparable QoE lab user measurement, they recommend to stay within this limit in order to achieve a good balance between results quantity and results quality. But the result can hardly be applied to the actual network environment because of the limitation of the simplicity of the experimental network. Yet it can provide a helpful guidance for QoE evaluation.

5.2.3 Measurement Under Simulation Network Environment

Besides the above two network environments, the simulation network environment is also chosen by many researchers. The advantage of the simulation network envi-

Fig. 5.3 Evalvid
architecture [16]

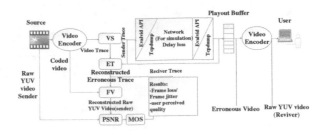

ronment is that the complex network can be set up quickly in order to meet the
need of complex experimental environment for researches. Liyan et al. used a NS-
3 simulation tool to build a simulation platform for LTE-based mobile streaming
network transmission and added an Evalvid tool to obtain the MOS values of users
[13–15]. Figure 5.3 presents the evaluation framework of Evalvid architecture depicts
the architecture of Evalvid [16]. As a result, they achieved a better fitting effect
whereby they obtained a single-index and multi-index evaluation model under LTE
network. Q. Zheng et al. used an open LTE simulator-OpenSim to provide a simulation of virtual LTE network with the ability to connect physical hosts through real
wired connection in real-time [17]. The related logical entities of video streaming in
transport and application layers can be deployed on remote hosts and will no longer
be limited by the simulator framework. However, due to the limitation of actual
situation, the simulation data is insufficient in some way, which results in the impre-
cision of the evaluation model. Although the experiment is easy to carry out, it lacks
parameters which consider the user effect. Therefore, to obtain more precise results,
the actual network data and parameters of users are necessary to be considered.

5.3   Design of Subjective Measurement

To quantify user QoE, MOS is usually used in subjective tests. It is user's feedback
after video watching so that it can reflect user subjective measurement on video
accurately. A great number of methods on QoE measurement focus on parameters in network layer (jitter, packet loss or delay) and application layer (resolution,
frame rate, etc.) [15]. But evaluation results based on those methods rarely consider
user-personal impact. As users increasingly participate in video streaming watching,
human-centric especially psychological factors weigh more in QoE measurement.
The experiment aims to clarify the relationship between technical factors, user preference and resultant user's QoE which is closely related to many factors. Instead
of objective factors, subjective factors play an important role on QoE measurement.
Those factors are generally obtained by special experimental application which sup-
ports user's participation and measurement. Therefore, we design a platform which
can help to collect user preference and relevant factors when they are watching online
videos.

5.3.1 QoE Related Factors

With the development of smart phones and mobile communication, mobile terminals are increasingly chosen as the platform for user to watch videos. Therefore, the experiment focuses on short-formed video QoE measurement on user's own mobile phones.

(A) Content Classification

The experiment system stands on the view of content service providers and provides distinct videos with diverse content types. Considering the time limitation of short videos, the content type is classified into six categories which are cartoon, movie, news, sport, MV, and entertainment.

(B) Influencing Factors

User's evaluation of video is determined with many factors which can be classified with objective factors and subjective factors. Objective factors depend on the condition of network and video sources. Traditionally, those factors can be obtained from technical approaches and evaluated as QoS metrics. As to subjective factors, they are crucial determinants for the evaluation of personal evaluation since they can reflect user perception directly. From objective aspect, the video is visually different from definition which is influenced by resolution and bitrate, as shown in Fig. 5.4. In this experiment, we control the resolution of video to 480 p and adjust different bitrate to achieve three levels of video quality. Because the experimental network is limited to local network, the network influence is not considered in this part. Compared with objective factors, subjective factors are more complex. With collection and analysis of those factors, the user-personal perception can be reflected. The subjective factors can be extended to user's psychological conditions such as preference and user's profile information such as user gender. Also, other environmental factors can make an influence on evaluation result. But in this experiment, we keep them as constant as possible. User perception can be estimated by preference factors directly. In this experiment, the preference factors are quantified with three levels. Table 5.1 shows the preference criteria for video content.

Also, user-personal characteristics may be the influencing factor. We collect user's gender information and their terminal information for further analysis about personalized difference.

Table 5.1 Preference level criteria

Preference level	Description
1	Dislike
2	Do not like nor dislike
3	Like

Fig. 5.4 Influencing factors

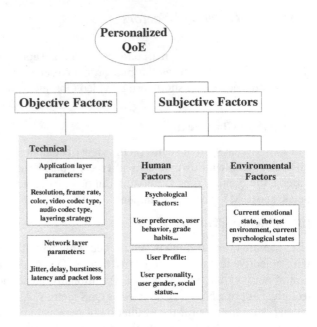

Table 5.2 MOS criteria

Score	Description
5	Imperceptible
4	Perceptible but not annoying
3	Slightly annoying
2	Annoying
1	Very annoying

(C) Measurement Criterion

According to ITU-T Recommendation P.910, MOS is chosen as the score criterion for QoE measurement. The score can reflect users' overall experience for the target video. Also, compared with other user's survey, it is more intuitional and clear to be collected. As a result, the experiment uses MOS as measurement criterion, as shown in Table 5.2.

5.4 Platform Infrastructure on Streaming Media Application Scenario

5.4.1 Supporting System Architecture

The architecture adopts B/S (Browser/Server) structure. Users watch videos on the browser of mobile terminals and submit feedbacks to the server. As shown in Fig. 5.5,

Fig. 5.5 The overview system architecture

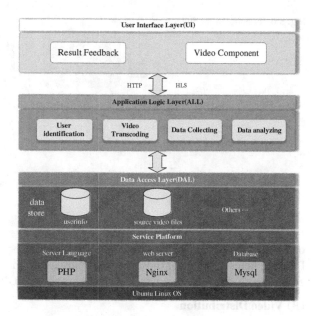

the whole server operates on Ubuntu Linux system which provides a supporting platform for the experiment system. The experimental system is supported by LNMP (Linux—Nginx—MySQL—PHP) which is a free and efficient web service system. Compared with LAMP (Linux—Apache—MySQL—PHP), LNMP differs from the web server. Nginx is a lightweight and high-performance web server and a reverse proxy server. Compared with other kinds of web servers, it has the advantages of less internal memory occupation and good performance on concurrence. As a result, it is high-performance on static and index file processing, which improves the efficiency for M3U8 file indexing to implement HTTP Live Streaming (HLS) protocol. Also, it is easy to configure, which releases the pressure for the system establishing.

Both video distribution and result feedback functions are built up on top of the support system architecture.

5.4.2 Functional Modules

From the functional aspect, the streaming media system can be divided into three parts, i.e., encoder, server, and terminal. The encoder is responsible for the content preparation function. The server is to provide platform to distribute prepared videos. With request and download, the terminal can play the video for users. As shown in Fig. 5.6.

(A) Content Preparation

FFmpeg, which is a frequently used multimedia open source tool, serves for the content preparation to complete encoder function. Original videos are encoded based

Fig. 5.6 Functional modules

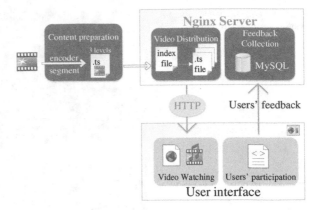

on MPEG-2 system layer standard and output with transport streaming format. Also, FFmpeg helps to segment those transport streaming into a series of sequential and equal-length small TS files which will be sent to video distribution module then.

(B) Video Distribution

For a conversation, the browser will download an extended M3U (m3u8) playlist which acts as an index to find streaming media segments. Video distribution module, which is supported by Nginx server, is to distribute the segmented TS files and related index files to client player. It is responsible to respond to users' requests and allocate prepared media streaming with HLS protocol.

(C) Terminal playing

With user's request, browsers obtain and download the index file which specifies the available TS files, the decryption key and the location of the other replacement flow on the server. When the media file buffer is enough large to support playing, the client will assemble them into a sequential TS streaming and send to the player.

(D) Feedback Collection

The video player embeds in the webpage which can be browsed on mobile terminals. Users watch videos on browser and submit their subjective data from web pages. Then, the form will be submitted to server to complete data collecting. In this experiment, users' preference, measurement scores, gender, video contents will be collected and stored in database.

5.5　Measurement Procedure

5.5.1　Crowdsourcing

Although lab based test tends to provide more accurate results, crowdsourcing seems to be an alternative way with time and cost limitation. Crowdsourcing is to invite

anonymous volunteers from the Internet to participate in subjective test. It widens
the range of testers and makes the measurement as simple as possible. The testers
in local area network are asked to watch online videos and then give a quality and
preference measurement for each one.

5.5.2 Measurement Description

During the experiment, the QoE evaluation results are collected with corresponding
users' subjective information. The relationship between the values of QoE parameters
and application metadata is analyzed and modeled accordingly. That is the collected
data should include the preference score, the quality measurement and the videos
with different bitrates. The video, whose resolution is 480 P, encoded with FFmpeg
into H264/AAC at three different bitrates: 128, 512, and 1024 kbps. Those prepared
videos are embedded in HTML webpages so that users can watch videos through
their browser on mobile terminals. Table 5.3 shows the information about videos
used in this experiment.

User Operations

Users use their mobile terminals to connect and login to the campus network.
Then, a given website is opened and a list of video is shown on the webpage (shown
in Fig. 5.7). The users can access any video randomly and then give a brief evaluation
for each.

5.5.3 Measurement Result

Users' feedbacks of a total of 18 videos assessment are stored in database. Based on
those crowdsourcing data, brief analysis has been conducted. As shown in Fig. 5.8,
it is observed the score gaps among the videos with different qualities. A tradeoff
between video quality and user video preference is also observed. That is, the users
are willing to tolerate poorer video quality when they are watching their favorite
videos.

Table 5.3 Video used in test

Item	Value
Content	Cartoon, MV, News, Sports, Movie, Entertainment
Amount of videos in each content	3
Resolution	480p
Bitrate	128, 512, 1024 kbps
Length of each video	1 min

Fig. 5.7 The video list on
the webpage

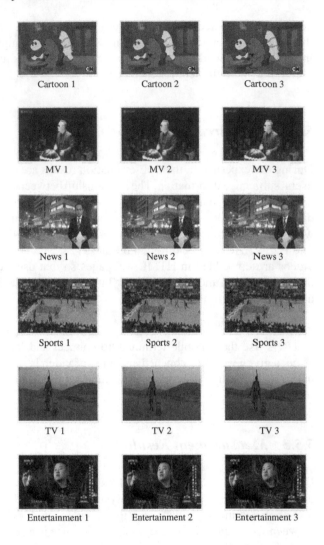

Figure 5.9 shows the average score for each content type with different quality videos. Compared with other content types, news seems to have a lower score on lower video quality degree and there is not a big score gap for cartoon among the different video qualities. For some specific content type, users pay more attention to the details especially facial detail, such as news, sports, etc. The bitrate change will have a greater impact on these content types. It is also observed that even with the same user preference and video quality, the QoE is still various for the current video application. It demonstrates that further research work should be done for QoE measurement.

Figure 5.10 illustrates that if a certain video is not preferred, user is quite sensitive to the video quality change.

Fig. 5.8 Average score of videos with different preference

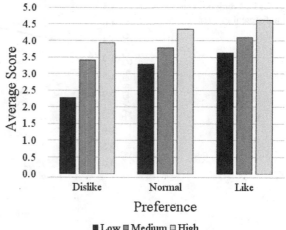

Fig. 5.9 Average score of videos with different bitrates

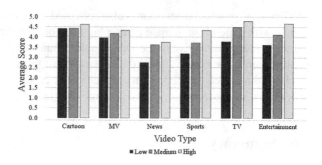

Fig. 5.10 Average score for videos with different quality and preference

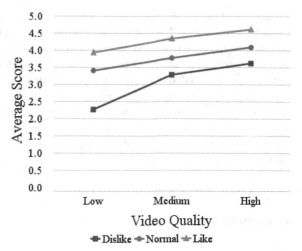

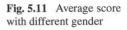

Fig. 5.11 Average score
with different gender

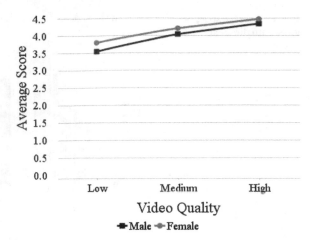

The users' gender information is also collected. As shown in Fig. 5.11, comparing female with male score, female gives a higher score for the video, which means that female is more insensitive to the video quality than male.

Based on those observations, the necessity of QoE measurement is validated. It can be summarized that the technical factors are crucial for the QoE measurement. In addition, user perception is also an important metric.

5.6 Summary

In this chapter, the QoE measurement platform for implementation and demonstration is presented and described. The measurement factors and criterion are given respectively. In order to clearly articulate the whole infrastructure, we use a streaming media application as an example. A detailed design of the subjective experiment is also given especially including the QoE-related factors and measurement criterion definition. A detailed supporting architecture is described and an experiment procedure is given. Procedure and infrastructure is then presented and illustrated. Some observations are also given on the influences of different factors such as user preferences, video qualities, video content type, and user gender information etc. Finally, a conclusion is given.

References

1. D. Schuurman, K. De Moor, L. De Marez, and T. Evens. "A Living Lab research approach for mobile TV". *Telematics and Informatics*, 28(4):271–282, Nov. 2011.
2. K. De Moor, I. Ketyko, W. Joseph, T. Deryckere, L. De Marez, L. Martens, and G. Verleye. "Proposed framework for evaluating quality of experience in a mobile, testbed-oriented living

lab setting". *Mobile Networks and Applications*, 15(3):378–391, Jan. 2010.

3. Miguel P. De L., M. Eriksson, S. Balasubramaniam, and W. Donnelly. "Creating a distributed mobile networking testbed environment-through the living labs approach". In *2nd International Conference on Testbeds and Research Infrastructures for the Development of Networks and Communities, 2006. TRIDENTCOM 2006*, pages 5–pp. IEEE, Mar. 2006.

4. V. Niitamo, S. Kulkki, M. Eriksson, and K. A. Hribernik. "State-of-the-art and good practice in the field of living labs". In *Proceedings of the 12th International Conference on Concurrent Enterprising: Innovative Products and Services through Collaborative Networks. Italy: Milan*, pages 26–28, 2006.

5. X. Yu, H. Chen, W. Zhao, and L. Xie. "No-reference QoE Prediction Model for Video Streaming Service in 3G Networks". In *2012 8th International Conference on Wireless Communications, Networking and Mobile Computing (WiCOM)*, pages 1–4. IEEE, Sep. 2012.

6. T. Kawano, K. Yamagishi, K. Watanabe, and J. Okamoto. "No reference video-quality-assessment model for video streaming services". In *2010 18th International Conference on Packet Video Workshop (PV)*, pages 158–164. IEEE, Dec. 2010.

7. T. De Pessemier, K. De Moor, A. Juan, W. Joseph, L. De Marez, and L. Martens. "Quantifying QoE of mobile video consumption in a real-life setting drawing on objective and subjective parameters". In *2011 IEEE International Symposium on Broadband Multimedia Systems and Broadcasting (BMSB)*, pages 1–6. IEEE, Jun. 2011.

8. Y. Huang, W. Zhou, and Y. Du. "Research on the user behavior-based QoE evaluation method for HTTP mobile streaming". In *2014 Ninth International Conference on Broadband and Wireless Computing, Communication and Applications (BWCCA)*, pages 47–51. IEEE, Nov. 2014.

9. S. Yu, W. Zhou, R. Tao, and Y. Hou. "Modeling for Short-Form HTTP Adaptive Streaming Considering Memory Effect". In *2015 10th International Conference on Broadband and Wireless Computing, Communication and Applications (BWCCA)*, pages 82–87. IEEE, Nov. 2015.

10. A. Sideris, E. Markakis, N. Zotos, E. Pallis, and C. Skianis. "MPEG-DASH users' QoE: The segment duration effect". In *2015 Seventh International Workshop on Quality of Multimedia Experience (QoMEX)*, pages 1–6. IEEE, May. 2015.

11. Y. Shen, Y. Liu, N. Qiao, L. Sang, and D. Yang. "QoE-based evaluation model on video streaming service quality". In *Globecom Workshops (GC Wkshps), 2012 IEEE*, pages 1314–1318. IEEE, Dec. 2012.

12. R. Schatz, S. Egger, and K. Masuch. "The impact of test duration on user fatigue and reliability of subjective quality ratings". *Journal of the Audio Engineering Society*, 60(1/2):63–73, Mar. 2012.

13. L. You, W. Zhou, Z. Chen, and W. Wu. "A novel method to calculate QoE-oriented dynamic weights of indicators for telecommunication service". In *TENCON 2013-2013 IEEE Region 10 Conference (31194)*, pages 1–4. IEEE, Oct. 2013.

14. M. Fiedler, T. Hossfeld, and P. Tran-Gia. "A generic quantitative relationship between quality of experience and quality of service". *Network, IEEE*, 24(2):36–41, Mar.-Apr. 2010.

15. H. Kim, D. H. Lee, J. Lee, K. Lee, W. Lyu, and S. Choi. "The QoE evaluation method through the QoS-QoE correlation model". In *Fourth International Conference on Networked Computing and Advanced Information Management, 2008. NCM'08*, volume 2, pages 719–725. IEEE, Sep. 2008.

16. J. Klaue, B. Rathke, and A. Wolisz. "Evalvid–A framework for video transmission and quality evaluation". In *Modelling techniques and tools, Computer performance evaluation*, pages 255–272. Springer, Sep. 2003.

17. Q. Zheng, H. Du, J. Li, W. Zhang, and Q. Li. "Open-LTE: An open LTE simulator for mobile video streaming". In *2014 IEEE International Conference on Multimedia and Expo Workshops (ICMEW)*, pages 1–2. IEEE, Jul. 2014.

Chapter 6
Conclusion

Abstract In this chapter, we summarize the main innovations and contributions of this book before future work is discussed.

6.1 Conclusion Remarks

On the basis of traditional radio resource management, we aim to develop a data-driven personalized radio resource management technology in this book, focusing on enhancing users' QoE. The user objective/subjective information is taken into account to guide wireless resource management accordingly. The main innovations and contributions include:

(1) A user's QoE description framework is established with comprehensive information of multi-dimensional context. A specific user tends to have a unique feeling on a particular application or service. This book provides a way to be able to utilize user personalized information when modeling QoE. Such personalized QoE model is expected to be more powerful.

(2) An architecture of Data-driven Personalized QoE Management is also proposed in this book, which consists of two modules: namely training module and control module. The training module is responsible for training and tuning the user preference prediction model. The control module collects the data about the users and provides the user preference instantly by using the model given by the training module. An efficient radio resource management process is done given the predicted results.

(3) This book tends to explore how to use data-driven methods for user preference prediction model. Two models are used in this book including Bayesian Graphic Model and Context Aware Matrix Factorization Model, which is borrowed from machine learning community. The experimental result demonstrates that the data-driven model is more robust and more promising in the era of big data.

(4) Resource allocation strategies based on QoE optimization are further studied in this book, where QoE-based resource allocation problem is investigated from both non-personalized and personalized aspects. A QoE-based resource allocation algorithm is proposed to attain suboptimal solution for QoE maximization problem, providing guidance to conduct QoE-aware radio resource management. Sim-

Y. Wang et al., *QoE Management in Wireless Networks*, SpringerBriefs
in Electrical and Computer Engineering, DOI 10.1007/978-3-319-42454-5_6

ulation results validate that a more refined resource allocation can be achieved by personalized QoE management.

(5) An experiment is conducted to collect QoE evaluation with corresponding user subjective information in the designed particular application scenario. With analysis on users' feedbacks, it can be summarized that these influencing factors are crucial determinants for the evaluation of QoE, and the effect of user preference on the perception is validated.

6.2 Future Work

Personalized QoE management in mobile Internet is a promising research topic now, which is also greatly significant for radio resource management. This book provides guidance on how to utilize available user data in personalized QoE management. Of course further research work should be concerned in the future studies including:

(1) More tests and experiments are necessary to validate the proposed data-driven model. More QoE influencing factors including age, gender, job, mood etc. should be involved. And better QoE assessment methods are preferred.

(2) In this book, the target application scenario is emphasized on the stream media service. Other types of services are under investigation, which possibly cause the new model structure, new influence factors and/or new assessment infrastructures.

(3) The objective function for QoE-oriented radio resource management is changeable with respect to the variant of applications, communication qualities user requirements, etc. A more resilient objective function is preferred and a set of more efficient and effective optimization methods has to be studied.

Printed in the United States
By Bookmasters